The Construction of Buildings

Volume 5

THIRD EDITION

THE CONSTRUCTION OF BUILDINGS

Volume 5

BUILDING SERVICES

THIRD EDITION

R. BARRY

Architect

WATER, ELECTRICITY AND GAS SUPPLIES
FOUL WATER DISCHARGE, REFUSE STORAGE

Blackwell
Science

DISTRIBUTORS

Marston Book Services Ltd
PO Box 269
Abingdon
Oxon OX14 4YN
(*Orders:* Tel: 01235 465500
           Fax: 01235 465555)

USA
Blackwell Science, Inc.
Commerce Place
350 Main Street
Malden, MA 02148 5018
(*Orders:* Tel: 800 759 6102
               781 388 8250
           Fax: 781 388 8255)

Canada
Login Brothers Book Company
324 Saulteaux Crescent
Winnipeg, Manitoba R3J 3T2
(*Orders:* Tel: 204 224-4068)

Australia
Blackwell Science Pty Ltd
54 University Street
Carlton, Victoria 3053
(*Orders:* Tel: 03 9347 0300
           Fax: 03 9347 5001)

A catalogue record for this title is available from
the British Library

ISBN 0-632-04932-4

Contents

Preface

Since publication of the first volume of *The Construction of Buildings* in 1958, the five volume series has been used by both lecturers and students of architecture, building and surveying, and by those seeking guidance for self-build housing and works of alteration and addition.

The layout of all volumes has remained the same until this latest revision to Volume 5. A wide right hand column of text has now been adopted to facilitate the inclusion of small diagrams in the left hand column, and larger diagrams within the text column so that, wherever possible, the diagram is adjacent to the relevant text for ease of reference. Bold subheadings in the left hand column provide a quick reference to the reader. Other volumes in the series will be revised to include the new layout.

Volume 5 deals with building services, water, electricity and gas supplies, foul water discharge and refuse storage. It does not attempt to cover heating, ventilating, air-conditioning, lifts or electronic or fire services.

The new Third Edition, the first revision to the volume in ten years, has been updated and includes changes to the 16th Edition of the Wiring Regulations and the current regulations on gas supply that concern safety.

1: Water Supply and Distribution

WATER SOURCES

History

Following the destruction of the Roman engineering works in Britain, there remained no organised piped supplies of water up to the twelfth century. In Saxon and Norman times the sole sources of water were streams, rivers, natural springs and man-made wells, on which the small population depended and around which it settled.

Monastic supplies

During the twelfth and thirteenth centuries the monasteries organised supplies of water usually from springs on high ground, from which water flowed by gravity, through wood, earthenware or lead pipes. Many of these monastic water systems were extended to supply neighbouring settlements. At the time water was less used by the general population outside the monasteries for either washing or drinking, than it is today.

With the dissolution of the monasteries in the sixteenth century the monastic water supplies were taken over by the civic authorities. The supply of water was generally in the hands of a local contractor who levied a charge for drawing water from the conduits and public fountains.

Water wheel

This rudimentary system of water supply was in existence until the end of the sixteenth century when the water wheel was first used to raise water to a cistern from which it flowed by gravity through wood or earthenware pipes to fountains, wash houses and buildings. The large wooden water wheels were driven by the river or stream from which they drew water. The wheel scooped up water, which was discharged at a high level, or drove air pumps which forced the water up.

Newcomen engine

From the middle of the eighteenth century the Newcomen steam engine gradually replaced the water wheel as the motive power to raise water. Water was piped under the head pressure from cisterns to communal conduits or fountains from which the populace drew their supply and from which water carriers collected water to retail for a small charge. A few town houses had a piped supply of water, the charge for which was beyond the means of the majority. The supply of water was, by and large, in the hands of commercial suppliers for profit, and this arrangement continued up to the middle of the nineteenth century.

The rapid increase in the urban population that followed the Industrial Revolution, culminated in the spread of cholera epidemics during the first half of the nineteenth century. The prime cause of cholera was gross pollution of water supplies by untreated sewage discharged directly into streams and rivers from which water supplies were drawn. Following investigations of the cause of cholera and reports of the total inadequacy of clean water supplies for the urban population, principally by Edwin Chadwick, the Waterworks Clauses Act was passed in 1847.

Waterworks Clauses Act

The Waterworks Clauses Act was the first comprehensive Act that standardised water work practice throughout the country. The Act controlled the construction of waterworks, laying of pipes, supply of water, fouling of water and obliged the water undertakers to supply water constantly, in sufficient quantity and at a reasonable pressure to all houses demanding water.

Prior to this Act the intermittent supply of water was a gravity supply, so that a storage cistern in each building was necessary to maintain a constant supply and those buildings above the level of the supply had no piped water. With a constant supply there was in theory no longer need for a storage cistern in each building and with reasonable pressure piped water could be provided to each floor level.

During the 130 years following the Waterworks Clauses Act municipalities largely took over the supply of water. The control of water pollution improved dramatically and water undertakings expanded the supply to the increasing urban population. The development of the supply and control of water supplies led to the Water Act 1973, which set up ten regional water authorities charged with the collection and supply of water in their area and the control of sewage treatment works and the pollution of water.

Quantity of water

There is no overall shortage of water in this country. In an average year rainfall over England and Wales is 900 mm, half of which is lost to evaporation and transpiration, leaving an average of 190 billion litres per day from which the public water supplies and industry take about 17 billion litres. Thus the direct supply consumption, including industry and agriculture, is about 10% of the potential supply available from rainwater.

The organisation of public water supply is not concerned, therefore, with the quantity of water available but with the collection and storage in order to cover seasonal variations and requirements, the movement of water from areas of surplus to areas of shortage and the overall control and treatment of effluents to avoid pollution and maintain the quality of water supplies.

Surface water

Surface rainwater that drains across lowland ground to rivers will contain various impurities from field drainage and factory and sewage effluents, whereas surface water that drains to upland rivers and lakes will be comparatively free of impurities.

Lowland surface water is stored in reservoirs in which some purification occurs naturally on exposure to air. Further purification is generally necessary by sedimentation and chemical additives.

In England and Wales a high proportion of water supplies is drawn from surface water.

Ground water

Rainwater that has percolated through permeable strata to a subsurface level will generally contain few impurities due to the filtration effect and may need little purification. Ground water that is drawn from deep wells and borings is usually free from impurities.

Storage

In lowland areas water is mainly drawn from rivers and stored in reservoirs against seasonal variations in supply and demand and in which the chemical and biological content of the water may be controlled. In upland areas, water is usually stored in natural or man-made lakes fed by run-off from the surrounding higher ground and in reservoirs fed by wells and boreholes.

Most urban and rural areas of this country are served by a supply of water piped from the water supplier's main. Water authorities are required by statute to provide a supply of wholesome water for which they may require a capital contribution towards the cost of running the supply pipe from their nearest main. In outlying areas it may be more economical to draw supply from a well or borehole than pay the capital contribution.

Wells, boreholes and springs

A well is a shaft sunk or excavated below the level of ground water or into permeable subsoil, water bearing strata. The shaft, usually circular, is lined with brick, stone or precast concrete sections to maintain the sides of the well.

A borehole is a steel-lined shaft driven or drilled into the ground to a water bearing stratum.

Springs break where the water level in a permeable stratum is above the level of the junction of permeable and impermeable strata as illustrated in Fig. 1.

Wells are defined as shallow and deep wells. A shallow well is one that is sunk to collect ground water and a deep well is one that is sunk to collect water from below the first impermeable stratum.

The distinction is made between shallow and deep wells in relation to the quality or purity of the water drawn from each. The shallow well may draw surface and ground water that could be contaminated whereas water from a deep well is less likely to be contaminated as the water has percolated to a permeable stratum and has been purified by

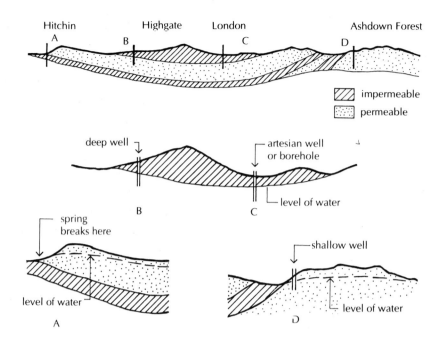

Fig. 1 Section across London Basin and Weald.

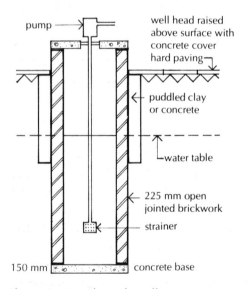

Fig. 2 Section through well.

Consumption

filtration through the over lying strata. Fig. 1 illustrates the difference between deep and shallow wells.

Wells are generally excavated by hand and the smallest diameter of the well is dictated by the space required by the excavator to work in the shaft. The well hole is lined with stone, brick or precast concrete rings. The lining of a shallow well should be rendered impermeable for some distance down from the surface with a surround of puddled clay or concrete to exclude surface water, that may be polluted, as illustrated in Fig. 2. The lining of a deep well should likewise be made impermeable down to the first impermeable stratum to exclude surface and ground water. Water is raised from wells by a pump (see Fig. 2). Shallow wells may dry up during summer months.

An artesian well is sunk, in a valley, to a permeable stratum from which water rises with force from the folded permeable layer sandwiched between impermeable layers as illustrated in Fig. 1.

A borehole is sunk by driving or drilling steel lining tubes down to a water bearing permeable stratum. Water enters the perforated or slotted end of the tube and is raised by a force pump.

Over the last 40 years the consumption of water in this country has doubled and is still rising. The water companies, the electricity generating industry and fish farming are the biggest users.

An average household uses about 380 litres every day of which 3% is used for drinking and cooking. A toilet flush takes about 7, a bath

80, a shower 35 and a washing machine 80 litres every cycle. A hosepipe or sprinkler can use 540 litres per hour, as much as the average person uses every four days.

Domestic water to most of the established urban settlements in this country is drawn from mains laid down some hundred or more years ago which by now, due to deterioration of old pipework, suffer considerable loss of water by leaks.

The combination of a naturally variable climate and periods of low rainfall have caused some recent water shortages. The remedy for this is a combination of more economical use of water, renewal of existing mains and works of feeder mains to move water from areas of high rainfall and low population to areas of drought and high consumption.

Water quality, purity and wholesomeness

The measure of the purity or wholesomeness of water is its freedom from pathogens which may be the cause of waterborne diseases in man. The waterborne diseases are often caused by the pollution of water by untreated or partly treated sewage.

It is not practical to test all the water that is used for traces of pathogens. Instead samples are taken for evidence of pollution by sewage as an indication of the likelihood of disease-carrying pathogens.

Water is purified by sedimentation and by filtration and by the addition of very small doses of chlorine to the water. The action of chlorine in water is disinfection whereby the chlorine reduces the infectious organisms to extremely low levels.

Acidity, alkalinity

Rainwater and water that contains decomposing organic matter such as peat, is acid water. Acid water is corrosive to iron and steel and will take lead into solution. As acid water is plumbo solvent, lead pipes should not be used in distributing acid water as lead is a cumulative poison.

Alkaline water contains calcium bicarbonate which is the constituent of temporary hardness. Pure water consists of hydrogen and oxygen and a small number of positive and negative ions. The measure of the acidity or alkalinity of water is pH, the letters used to denote the concentration of hydrogen ions. An acid water has a pH value of less than 7.0 and alkaline water a pH above 7.0.

Hard water, soft water

Much of the water drawn from underground sources such as chalk and limestone beds and from surface water on and in clay deposits, is said to be hard. Free carbon dioxide in underground water combines with chalk and limestone to form calcium bicarbonate, which is soluble in water (temporary hardness) and calcium and magnesium sulphate and chloride (permanent hardness).

Temporary hardness, permanent hardness

Table 1 Water hardness.

0 to 50 mg/l	Soft
50 to 100 mg/l	Moderately soft
100 to 150 mg/l	Slightly soft
150 to 200 mg/l	Moderately hard
over 200 mg/l	Hard
over 300 mg/l	Very hard

Much of the water drawn from underground sources such as chalk and limestone beds and from surface water in clay deposits, is said to be hard. Free carbon dioxide in underground water combines with chalk or limestone to form calcium bicarbonate, which is soluble in water (temporary hardness) and calcium and magnesium sulphate and chloride (permanent hardness).

When soap is dissolved in hard water it reacts with the minerals in the water to form a scum on the surface before it combines with water to form a lather. The excessive rubbing required to form a lather together with the scum formed is the reason the water is described as hard. More soap is required to form a lather in hard water than in soft water. It is advantageous therefore to treat hard water to reduce the hardness for washing purposes.

Table 1 indicates the degree of hardness of water. Hardness is measured in mg/litre as $CaCO_3$ (calcium carbonate). The range of hardness acceptable for general washing purposes lies between 100 and 200 mg/litres.

Drinking water

For drinking many prefer the taste or palatability of soft water from deep wells and fast flowing streams, while others prefer the taste of a hard water. The temperature of the water affects the subjective judgement of taste: the colder the water the more palatable it is said to be. Most will agree that stagnant water and especially tepid stagnant water has an unpleasant taste. Obviously, there is no generally accepted measure of taste.

Water for washing

The constituents of soap disperse oil, grease and dirt in water to form a lather of minute bubbles that serve as a lubricant and assist in washing. The more readily soap combines with water to form a lather, the softer the water feels to the touch when washing. Water that readily forms a lather with soap is said to be soft and water that does not is said to be hard. The words soft and hard are used subjectively to express the common sense feel of water when used with soap for washing. The descriptions hard and soft have been adopted to define the mineral content of water.

Scale, limescale

When water is boiled the soluble mineral bicarbonates form a hard scale, limescale, inside kettles and pans, and inside hot water and heating boilers so that progressively more fuel is required to heat the water, thus wasting energy. In time, heating and hot water pipes may become blocked by the build-up of scale. The harder the water and the higher the temperature the greater the build-up of scale. There is, therefore, economic advantage in controlling the hardness of water.

Temporary hardness, hardness

Because the mineral bicarbonate hardness of water is converted to scale by heating, it is described as 'temporary hardness' or more accurately 'carbonate hardness'. The noncarbonate mineral content of water, that is unaffected by heating, is termed 'permanent hardness' or 'noncarbonate hardness'. Bicarbonates of calcium and magnesium dissolved in water cause temporary hardness, and sulphates and chlorides cause permanent hardness.

Water treatment

Water is usually treated to change a hard water to a soft or softer water to reduce scale formation and facilitate washing. Two methods are used: in the first, lime or lime with soda is added to the water, which brings about changes to the hardness compounds so that they become insoluble and precipitate by settlement or in filters, and in the second the nature of the hardness is changed in a base exchange softener. In the first method the hardness compounds are changed and removed, and in the second the hardness compounds are changed and remain in the water. In the second method of softening a natural or synthetic zeolite is used to convert compounds of calcium and magnesium to sodium carbonate-bicarbonate and sulphates that do not cause hardness. For domestic treatment small base exchange equipment is available.

Water purification

Water purification combines storage in reservoirs to allow suspended matter to settle, followed by filtration to remove both suspended and dissolved matter and a final treatment by chlorination.

WATER SUPPLY

Water mains

Water is supplied by the 'statutory water undertaker' required to supply a constant, potable (drinkable) supply of water for which service either a water rate is levied for domestic consumption or a charge by meter for most other users. The water rate is assessed on the rateable value of premises and charged as a percentage of the value either half-yearly or annually. This method of assessment for the use of water is currently under review.

Metered supplies are charged at the current rate of the consumption recorded.

Water meter

In many European countries the supply of water to all buildings is measured by meter and the cost charged by units of consumption. The advantage of making a charge through units of consumption is that it tends to encourage economy of consumption by the consumer who has a financial interest in maintaining his water installation. The disadvantage of metered supplies to the water authorities lies in the labour costs of periodic reading of meters and accounting for the actual consumption. It is open to any consumer in England to ask for a metered supply of water, providing the consumer bears the cost of fitting the meter.

Water is supplied, under pressure, through pipes laid under streets, roads or pavements. Cast iron, ductile iron, steel, concrete and more recently plastic pipes are used. In urban areas duplicate trunk mains feed street mains. By closing valves individual lengths of main may be isolated for repairs and renewals without interrupting the supply.

Connections to water main
Service pipe

Connections to the existing main are made by the water undertaker. The house or building service pipe connection is made to the main and the service pipe is run to a stop valve near to the site boundary of the building to be served. The stop valve is situated either immediately outside or inside the boundary. The purpose of the stop valve is to enable the water undertaker to disconnect the water supply where there is a waste of water in the building served, or non-payment of water rate or charge.

Supply pipe

The pipe that is run from the stop valve to and into the building is termed a supply pipe. The supply pipe is run underground and into the building as illustrated in Fig. 3. For convenience it is usual to run the supply pipe into the building through drain pipes to facilitate renewal of the pipe if need be.

It is the responsibility of the consumer to maintain so much of the incoming service pipe as is on his land. At the point that the supply pipe enters the building there should be a stop valve (see Fig. 3), to disconnect the supply for repair and maintenance purposes.

To reduce the risk of freezing, the supply pipe should be laid at least 750 mm below the finished ground surface and if the supply pipe enters the building and rises closer than 750 mm to the outside face of a wall, it should be insulated from where it enters the building and up to the level of the ground floor.

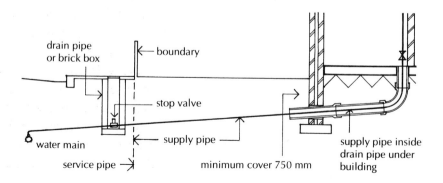

Fig. 3 Service pipe and supply pipe.

HOT AND COLD WATER SUPPLIES

Cold water supply

The intermittent supply of water that was common from the middle of the eighteenth to the middle of the twentieth century necessitated the use of a water storage cistern fixed at high level in each building to maintain a constant supply of cold water. These cisterns were designed to contain one or more days' use of water in the building, to allow for interruption in the supply.

Since those days, most water undertakers in England have required consumers to install a water storage cistern for the supply of cold water in each building, even though the water authority is obliged by statute to provide a constant supply. The cistern usually provides storage for twelve to twenty-four hours' consumption.

The principal reason for continuing the requirement for cistern-fed cold water supplies was that the airgap between the discharge of the supply pipe and the water level in the cistern acted as an effective break against backflow contamination of mains supply. In most European countries the use of cold water storage cisterns has long since been abandoned in favour of cold water supplied directly under pressure from the main supply on the grounds that roof level cisterns are 'expensive, unhygienic and unnecessary'. In 1988, for the first time in England, the new Water Supply Byelaws permitted the use of mains pressure supply of water as an alternative to the traditional cistern-fed supply.

Water Supply Byelaws

In the current byelaws there is a definition of the terms 'supply pipe' and 'distributing pipe'. A supply pipe is any pipe, maintained by the consumer, that is subject to water pressure from the authorities' mains, and a distributing pipe is any pipe (other than an overflow or flush pipe) that conveys water from a storage cistern or from hot water apparatus supplied from a feed cistern and under pressure from that cistern.

The Water Supply Byelaws set out requirements to prevent waste of water and contamination of mains supply water by backflow contamination.

Prevention of waste of water

Prevention of waste of water is mainly concerned with the overflow of roof level cisterns and cisterns to WCs and bidets and the unnecessary use of water through faulty taps and installations. The byelaws can exercise no control over the waste of water through taps carelessly left running, which is a cause of excessive consumption.

At present most consumers pay a charge for water through a water rate that is levied annually as a percentage of rateable value, taking no account of consumption. Where the charge for water is based on consumption measured by a meter, the consumer has a direct interest in the prevention of waste, through his pocket. A more sensible approach to prevention of waste would be through metered supplies as is the practice in most continental European countries.

Prevention of contamination

The principal concern of the byelaws is the prevention of contamination of mains supplied water by the flow of potentially polluted water from a supply or distributing system back into the mains water supply.

Flow of water from a supply or distributing system in a building, back into the mains supply, can occur when there is loss of pressure in the mains due to failure of pumps, or repair and maintenance on the mains, and also where a pumped supply in a building creates pressure greater than that in the mains.

Backflow

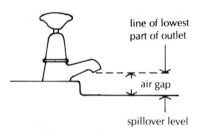

Fig. 4 Air gap to taps to fittings.

Loss or reduction of pressure in the mains supply could allow back-flow from fittings, through draw-off taps, flushing cisterns and washing machines. The concern here is that if, for example, a bath were to be overfilled and the pressure in the mains supply pipe reduced, there would be a possibility that polluted water might find its way back into the mains pressure supply system.

As a guard against this, all draw-off taps to baths, wash basins and sinks must either be fixed so that there is an air gap between the spill-over level of the fittings and the outlet of the tap, as illustrated in Fig. 4, or a double check valve assembly must be fitted to the supply pipe to each draw-off tap. These requirements apply equally to taps connected to mains pressure supply pipes and to distributing pipes from a cistern.

COLD WATER SUPPLY SYSTEMS

The two systems of cold water supply that are used are cistern feed and mains pressure.

Cistern feed (indirect)

The cistern feed supply is also termed the indirect system, because the cold water supply comes independent of the mains, from a cold water storage cistern.

Mains pressure (direct)

The mains pressure supply is also termed the direct system because the cold supply comes directly from the mains.

The advantages and disadvantages of the systems are:

Cistern feed – indirect supply
Advantages
(1)   The reserve of water in the cistern that may be called on against interruption of supply to provide regular flow.

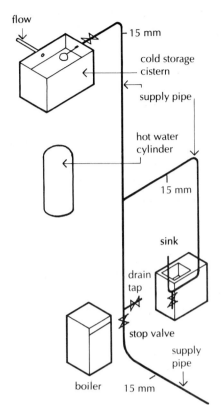

flow

15 mm

cold storage
cistern

supply pipe

hot water
cylinder

15 mm

sink

drain
tap

stop valve

supply
pipe

boiler   15 mm

Fig. 5 Cistern feed cold water supply.

(2) The air gap between the supply pipe and the water level in the cistern acts as an effective barrier to backflow into the mains supply, that can cause contamination.

Disadvantages

(1) The considerable weight of a filled cistern to be supported at high level

(2) The inconvenience of access to the cistern for inspection and maintenance

(3) The possibility of the cistern overflowing

(4) The need to insulate effectively the cistern and its associated pipework against freezing

Mains pressure – direct supply

Advantages

(1) A uniformly high pressure supply to all hot and cold water outlets some distance below the pumped head of the supply

Disadvantages

(1) Discontinuity of supply to all hot and cold water outlets if mains supply is interrupted

(2) The need for comparatively frequent inspection, maintenance and repair of many valves and controls to the system

There is little, if any, difference in the necessary capital outlay between a cistern gravity feed and a mains pressure supply.

CISTERN FEED COLD WATER SUPPLY

Fig. 5 illustrates the cold water supply to a two storey house. The supply pipe rises through a stop valve and draw-off or drain tap to the ball valve that fills the cistern. A stop valve is fitted close to the cistern to shut off the supply for maintenance and repairs.

Prior to the implementation of the 1988 Water Supply Byelaws it was a requirement that only one supply pipe connection be made to the kitchen sink as a drinking water outlet. The rationale for this was that roof space storage cisterns were often left uncovered and the water in the cistern became somewhat fouled with dust and other debris to the extent that the water in it was made unpalatable. Today with the use of covered plastic cisterns the water drawn from them is generally considered fit for drinking.

Cisterns

Galvanised steel cisterns

A cistern is a liquid storage container which is open to the air and in which the liquid is at normal atmospheric pressure. The traditional cold water storage cistern is manufactured from mild steel plate or sheet, welded or riveted together and galvanised after manufacture. Today the majority of smaller cisterns are moulded from plastic.

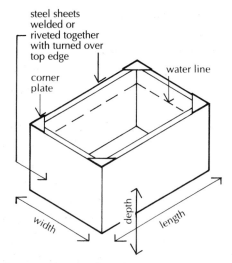

Fig. 6 Galvanised steel cistern.

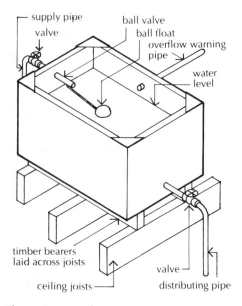

Fig. 7 Support for cistern.

Plastic cisterns

Fig. 6 is an illustration of a typical galvanised steel cistern. The cistern has turn-over flanged or angle section stiffeners and corner plates to provide some rigidity to the sides. A loose steel cover can be supplied to give some protection from contamination by dust and dirt. Cisterns that serve to supply drinking water have a close fitting top with a screened air inlet.

The useful life of these cisterns depends on the thickness of the galvanising coating, which is applied after manufacture. Thinly coated cisterns may rust after twenty to thirty years and need replacement. Those more thickly coated may last for the normal life of a building.

When a galvanised steel cistern fails, due to rusting, it is usual practice to replace it with two smaller cisterns to avoid enlarging trap doors in ceilings for access to roof spaces.

Small capacity galvanised steel cisterns are used as header or feed tanks to provide the necessary head of cold water supply to boilers where an open vented pipework system is used. The header or feed cistern or tank is fixed in a roof space or in some position above the boiler to provide the necessary head of water. An expansion pipe runs up from the boiler, to discharge over the header tank, so that water under pressure from overheating may discharge into the header tank.

Because of the weight of water they are designed to contain, all cisterns must be adequately supported on timber or steel spreaders to transfer the load over a sufficient area of timber or concrete ceilings and roofs, as illustrated in Fig. 7.

Galvanised steel cisterns with a capacity of 18 to 3364 litres are available in sizes from 457 to 2438 mm long, 305 to 1524 mm wide and 305 to 1219 mm deep. The small cisterns are for use as header tanks and the larger for such buildings as multi-storey flats, where a gravity feed cold water supply is used.

Stainless steel cisterns, manufactured from stainless steel sheets welded together, are used for drinking water storage where separate drinking water outlets are used. These expensive cisterns are used because of their freedom from rust and their durability.

Of recent years plastic cold water storage cisterns have been used instead of the traditional galvanised steel cistern. The advantages of these cisterns are that they do not deteriorate by rusting, have smooth surfaces and are comparatively lightweight for handling. The disadvantage is that the thin plastic from which these cisterns are made has poor mechanical strength in supporting the weight of the water they contain. Square-sided plastic cisterns are moulded with ribbed sides, and round section cisterns with a flanged top and the support of

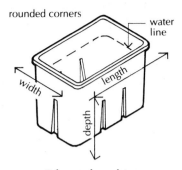

Polypropylene cistern

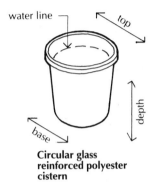

Circular glass
reinforced polyester
cistern

Fig. 8 Plastic cisterns.

a tightly fitting flanged cover as stiffening and support, as illustrated in Fig. 8.

An advantage of the round section plastic cisterns is that the material is sufficiently pliable to enable the empty cistern to be folded to an extent, to pass it through the small access trap door in ceilings as replacement for a defective cistern.

Rectangular section plastic cisterns are available with capacities from 18 to 227 litres, lengths of 430 to 1055 mm, widths of 305 to 635 mm and depths of 305 to 584 mm. Round section plastic cisterns are available with capacities from 18 to 227 litres, depths of 254 to 610 mm, base of 419 to 750 mm and tops of 464 to 845 mm.

The most convenient place for a cold water storage cistern is in a roof space below a pitched roof, or in a tank room on a flat roof or at some high level. In whichever position the cistern is fixed, there must be space around and over the cistern for maintenance or replacement of the ball valve.

The cold water storage cistern must be fixed not less than 2 m above the highest fitting it is to supply with water. This distance represents the least head of water necessary to provide an adequate flow.

The most convenient place for a cistern is therefore in the space below a pitched roof, or in a tank room or chamber above a flat roof or at some high level below the roof. Where a cistern is fitted inside a roof space it is necessary to spread the considerable weight of a full cistern across several ceiling joists.

The capacity of cold water storage cisterns is usually 114 litres per dwelling where the cistern supplies cold water outlets and WC cisterns, and 227 litres per dwelling where the cistern supplies both cold water to outlets and domestic hot water cylinders. The alternative is to calculate the capacity of the storage cistern at 90 litres per resident.

Cisterns are holed for the connection of the cold water supply, the distributing pipes to cold water outlets, and for the pipe to a hot water cylinder and the overflow pipe.

Ball valve, float valve

The water supply to the cistern is controlled by a ball valve that is fixed to the cistern, above the water line, and connected to the cold supply through a valve as illustrated in Fig. 7. A hollow copper or plastic ball, fixed to an arm, floats on the water in the cistern. As water is drawn from the cistern the ball falls and the arm activates the ball valve that opens to let water into the cistern. When the water has risen to the water line, marked on the cistern wall, the ball and arm rise to close the valve.

The two types of valve in use are the original Portsmouth valve and the diaphragm valve that is used for most new installations.

Portsmouth valve

The Portsmouth valve consists of a brass case and seating, as illustrated in Fig. 9. A brass piston, which is free to move inside the casing, is operated by a cam on the end of the ball arm that pivots to move the piston horizontally. To act as a water seal to the valve, a rubber or plastic washer is fitted to the end of the piston. As the water level in the cistern rises, the washer is pushed against the seating to shut off the water supply, which flows when the ball falls. An outlet allows water to discharge over the level of water in the cistern. The air space between the outlet and the water line of the cistern serves as a barrier to backflow contamination of the supply.

The original Portsmouth valve, which was noisy in operation, has been refined so that the piston is under equal water pressure at both ends, to reduce the noise of operation.

To reduce the noise of water falling from the valve outlet into an emptied cistern, it has been practice to fit a silencing tube to the outlet, to discharge water below the lowest level of water in the cistern. This is in contravention of Water Byelaw regulations as there is no longer an air gap as check to contamination.

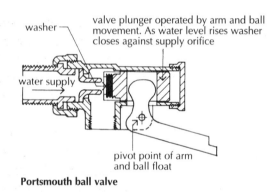

Portsmouth ball valve

Fig. 9 Ball valves.

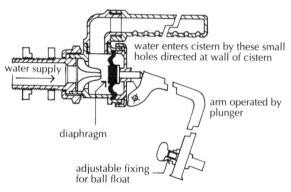

Diaphragm ball valve

Diaphragm valve

The diaphragm ball valve, illustrated in Fig. 9, consists of a brass case with outlet attachment in which slots direct water towards the walls of the cistern to reduce the noise of filling. A ball float and arm operates through a pivot to move a rubber diaphragm up to and away from a nylon nozzle, to shut and open the water supply.

This valve, which operates quietly and efficiently, will need only occasional attention to clean the diaphragm of particles so it makes an effective seal with the nozzle.

In general, ball valves operate efficiently and require infrequent maintenance to renew the washer of the Portsmouth valve and clean the diaphragm. Against the possibility of the cistern overflowing due to a faulty valve, an overflow pipe is fitted to the cistern above the waterline and carried out of an external wall.

Air gap to ball valve

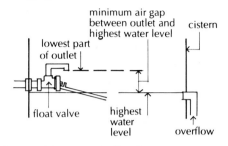

Fig. 10 Air gap to ball valve.

To prevent contamination of mains supply of water by back-siphonage, backflow or cross connection, the Water Supply Byelaws require an air gap between the outlet of float valves to cisterns and the highest water level in the cistern. This air gap is related to the bore of the supply pipe or the outlet and is taken as the minimum distance between the outlet of the float valve and the highest water level when the overflow pipe is passing the maximum rate of inflow to the cistern. Fig. 10 is an illustration of the air gap necessary to float valves to cisterns.

Overflow warning pipe

As a precaution against failure of the valve, and consequent overflow of the cistern, most water undertakers require an overflow warning pipe to be connected to the cistern above the water line and carried out of the building to discharge where the overflow of water will give obvious warning. The overflow pipe should be of larger bore than the supply pipe to the cistern and preferably twice the bore of the supply pipe and not less than 19 mm bore.

Insulation

The Water Supply Byelaws include a requirement that water storage cisterns be insulated and fitted with a close fitting cover that excludes light and insects and is not airtight, and that the cistern shall be adequately supported to avoid distortion and be in a position where it may be readily inspected and cleaned, and valves readily installed, renewed or adjusted.

To meet these requirements it will be necessary to fit cisterns on a platform to support the cistern and spread its load to ceiling or roof rafters. The cistern should be surrounded with adequate insulation in the form of quilted or board insulation, strapped or otherwise securely fixed. The thickness of the insulation is not specified as it will depend on the position in which this cistern is fixed. Cisterns which are fixed inside the roof space under pitched roofs, where the insulation is at ceiling level and the roof space ventilated to minimise condensation, will require heavy, airtight insulation against freezing. The cover to the cistern will have to be insulated as will the base of the cistern where it is not in direct contact with some other insulation.

Fig. 11 is an illustration of a cistern with cover and insulation. The air inlet shown in the airtight cover has an insect screen, as does the overflow pipe.

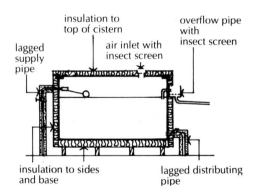

Fig. 11 Insulation to cisterns.

Cistern feed cold water distributing pipe system

Fig. 12 illustrates the cold water distributing pipe system for a two-storey house. The distributing pipe is connected to the cistern some 50 mm above the bottom of the cistern to prevent any sediment that may have collected from entering the pipe. A stop valve is fitted to the

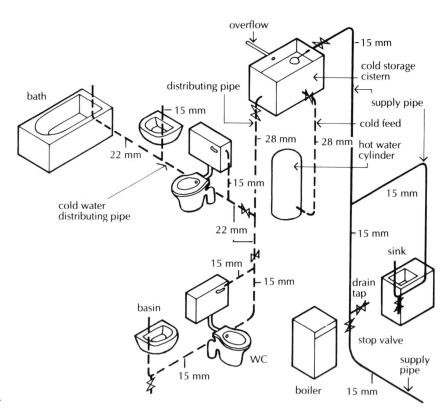

Fig. 12 Cold water distributing pipe system.

pipe adjacent to the cistern, isolating the whole system from the cistern in the event of repairs and renewals. The distributing pipe is carried down inside the building with horizontal branches to the first-floor and ground-floor fittings, as shown in Fig. 12.

The aim in the layout of the pipework is economy in the length of pipe runs, and on this depends a sensible layout of sanitary fittings. In Fig. 12 it will be seen that one horizontal branch serves both bath, basin and WC. For rewashering taps, stop valves are fitted to branches as shown. Where one branch serves three fittings as shown in Fig. 12, one stop valve will serve to isolate all three fittings.

Drain or draw-off taps should be provided where pipework cannot be drained to taps so that the whole distributing system may be drained for renewal or repair of pipework or when a building is left empty and the water in the system might otherwise freeze and fracture pipework or joints.

CISTERN FEED HOT WATER SUPPLY

Cistern feed hot water distributing pipe system

Fig. 13 illustrates the hot water distributing pipe system for a two-storey house. The hot water is drawn from a cylinder which is fed by cold water drawn from the cold water storage cistern in the roof.

The cold water in the hot water cylinder is heated by a heat exchanger in the cylinder through which hot water circulates from the boiler. The cold water feed to the cylinder is run through a stop valve to the bottom of the cylinder.

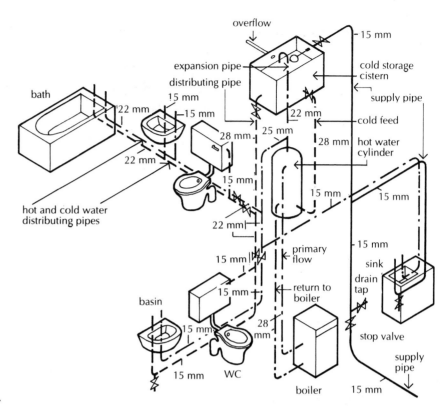

Fig. 13 Hot water distributing pipe system.

Hot water storage cylinder

Hot water storage cylinders are designed to contain water under pressure of the head of water from the cold water storage cistern. Most hot water storage containers are cylindrical and are fixed vertically to encourage cold water fed into the lower part of the cylinder to rise, as it is heated by the heat exchanger, to the top of the cylinder from which hot water is drawn, and so minimise mixing of cold and hot water.

The cold feed pipe to the cylinder is run from the cold water storage cistern and connected through an isolating stop valve to the base of the cylinder. The hot water distribution pipe is run from the top of the cylinder to the draw-off branches to sanitary appliances and is carried up to discharge over the cold cistern as an expansion pipe, in case of overheating, as illustrated in Fig. 13.

Vented hot water system

Because the expansion pipe discharges through the open end of the pipe, over the cold water storage cistern, in case of overheating and expansion of the hot water, this arrangement is described as a vented hot water system and the cylinder as a vented or open vented system. This is to distinguish it from the unvented system used with mains pressure system.

The required storage capacity of the cylinder depends on the number of sanitary appliances to be served and the estimated demand. A limited experiment by the Building Research Station suggests that the average consumption of domestic hot water is in the region of 50 litres per person per day. The interval between times of maximum demand on domestic hot water are longer than the recovery period required to reheat water in storage systems, and it is reasonable, therefore, to provide hot water storage capacity of 50 to 60 litres per person.

Storage cylinders are made either of galvanised sheet steel or copper sheet welded or riveted. Fig. 14 is an illustration of typical hot water storage cylinders. In course of time galvanised steel cylinders may rust and their average life is about 20 years, whereas a copper cylinder may have an unlimited life. Which of these two is used will depend on the pressure of the head of cold water, the nature of the water and the pipework used, and initial cost consideration. Most steel cylinders can support greater water pressure than copper cylinders, and steel cylinders are appreciably cheaper than copper cylinders.

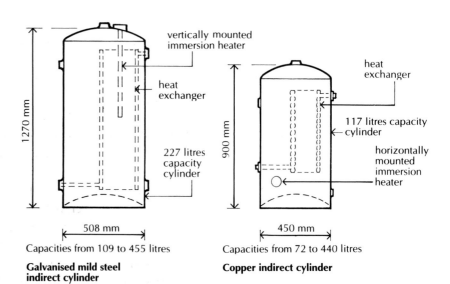

Fig. 14 Indirect hot water storage cylinders.

Indirect cylinder

The hot water storage cylinders illustrated in Fig. 14 are indirect cylinders, so called because the primary hot water from the boiler exchanges its heat indirectly through a heat exchanger to the hot water supply, there being no connection between the water from the boiler and the hot water supply. The purpose of this indirect transfer of heat is to avoid drawing hot water directly from the water system of the boiler. Where hot water is drawn directly from a boiler it has to be replaced and in hard water areas each fresh charge of water will deposit scale inside the boiler and its pipework and in time the build up of scale will reduce the efficiency of the boiler and the bore to its pipes. With an indirect cylinder there is no replacement of water to the boiler and its primary pipes and therefore no build up of scale.

Scale formation is proportional to water temperature. There is less build up of scale in the secondary hot water circulation because of the lower water temperature in the system. Indirect cylinders are in addition a protection against the possibility of drawing scalding water directly from the boiler. For these reasons indirect cylinders are used.

Heat exchanger

The heat exchanger which is fixed inside the cylinder and immersed in the water to be heated, takes the form of a coil of pipes or annulus designed to provide the maximum surface area for heat exchange.

Primary flow

The system of pipes that carries hot water from the boiler through the heat exchanger and back to the boiler, is described as a primary flow system.

Hot water boiler or heater

For a small building such as the two-storey house illustrated in Fig. 13 it is general practice to utilise one boiler for both space heating and hot water, to economise in the initial outlay on heating equipment and pipework and to make maximum use of floor space.

The boiler, whether combined space heating and hot water or separate, may be fired by solid fuel, gas, oil or electricity, in that order of current costs, solid fuel being the cheapest. Solid fuel suffers the disadvantage of being bulky to store, and residual ash and clinker are time-consuming and dirty to clear. Oil requires a bulky storage container and tends to smell, while gas is the most convenient of the fuels. Electricity is less used because of its cost.

The temperature of the water heated by the boiler is controlled by a thermostat, which can be set by hand to a range of water temperatures from 55°C to 85°C, the boiler firing and cutting out as the water temperature falls and then rises to the preselected temperature.

Water heated in the boiler rises in the primary flow pipe to the heat exchange coil or container inside the hot water storage cylinder and as it exchanges its heat through the exchanger to the water in the cylinder it cools and returns through the primary return pipe back to the boiler for reheating. There is a gravity circulation of water in the primary pipe system. The primary flow and return pipes should be as short as practicable, that is the cylinder should be near the boiler to minimise loss of heat from the pipes.

The circulating pipes from the boiler through the heat exchanger are termed primary flow and return as they convey the primary source of heat in the hot water system.

Small, compact, so-called, 'space saving' boilers require a forced flow of water around the primary circulation system by an electrically operated pump. A small feed cistern (not shown in Fig. 13) provides the head of water required for the boiler and its pipe system.

Immersion heater

An electric immersion heater is fitted to the hot water storage cylinder to provide hot water when the boiler is not used for space heating.

An electric immersion heater is both an inefficient and an expensive means of heating water. Because of the comparatively small surface area of the immersion heater, it requires some four hours to heat the water in the cylinder as compared to two hours for the heat exchange coil from the boiler. This slow recharge rate is an inconvenience added to the high cost of electricity.

MAINS PRESSURE COLD WATER SUPPLY

Fig. 15 illustrates the cold water, mains pressure pipe system to a two-storey house. Here the first line of defence against contamination of the mains supply is a double checkvalve assembly fitted upstream of the stop valve as the supply pipe enters the building.

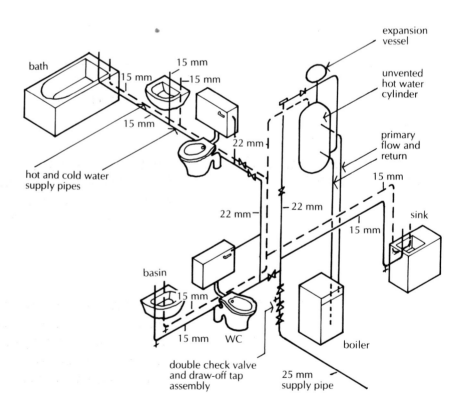

Fig. 15 Cold and hot water supply.

Check valve, non-return valve

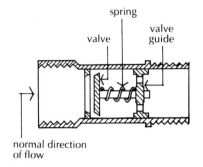

Fig. 16 Spring loaded check valve, non-return valve.

A check valve, more commonly known as a non-return valve, is a simple, spring-loaded valve that is designed to open one way to the pressure of water against the valve and to close when that pressure is reduced or stops. It therefore acts as a one way valve which closes against the direction of normal flow and prevents backflow from the supply pipe system into the mains. Fig. 16 is an illustration of a typical check valve.

A double check valve assembly is a combination of two check valves with a test cock between them. A test cock is a simple shut off device with a solid plug that, when rotated through 90°, either opens or closes.

In use, a check valve may become coated with sediment, particularly in hard water areas, and not operate as it should. To test that the check valves are working and acting as non-return valves, the stop valve is closed to test one check valve and opened to test the other with the test cock open.

Vacuum breaker

The new Water Supply Byelaws will accept a check valve and vacuum breaker, a pipe interrupter or other fittings or arrangement of fittings designed to prevent backflow, as an alternative to a double check valve assembly.

From the double check valve assembly the mains pressure supply pipe rises and branches horizontally to supply fittings on each floor level with stop valves to isolate sections for repair or maintenance and drain plugs, in an arrangement similar to that for the cistern-fed distribution pipe system. The supply pipe also supplies the unvented hot water storage cylinder.

MAINS PRESSURE HOT WATER SUPPLY

The hot water supply system for a small two storey house is illustrated in Fig. 17. From the double check valve assembly the supply pipe rises to fill the unvented hot water cylinder with cold water. The cold water in the cylinder is heated by a heat exchanger that exchanges heat from the boiler.

Unvented hot water cylinder

The difference between the traditional vented hot water system and the unvented system is that in the vented system the expansion of hot water is accommodated by the vented expansion pipe that will discharge an excess of expansion water to the cistern, and in the unvented system expansion of hot water is relieved by an expansion

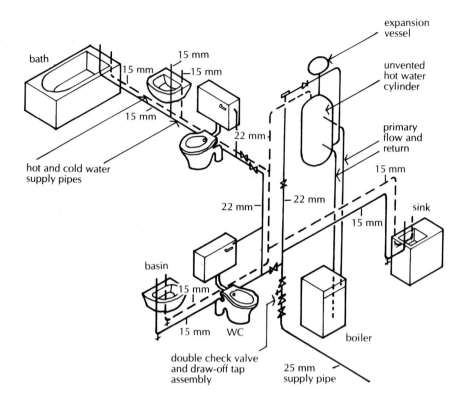

Fig. 17 Unvented hot water supply.

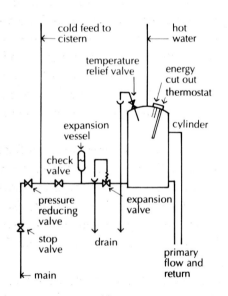

Fig. 18 Diagram of low pressure unvented hot water system.

Expansion vessel

vessel which contains a cushion of gas or air sufficient to take up the expansion by compression of the gas.

The advantages of the unvented over the vented hot water system, for the user, lie in the improved flow rates from showers and taps, reduction in noise caused by the filling of storage cisterns, and virtually no risk of frost damage. There is little if any economic advantage in the use of the unvented system. The saving in eliminating the cistern, feed and expansion pipes is offset by the additional cost of the expansion vessel and temperature and expansion control valves, and the necessary, comparatively frequent, maintenance of these controls.

Fig. 18 is a diagram of a low pressure unvented hot water cylinder. The pressure reducing valve is fitted to the feed pipe where low pressure systems are used and is provided to reduce mains pressure to a level that the cylinder can safely withstand. Where high pressure systems are used and the cylinder is designed to stand high pressure, the pressure relief valve is omitted.

An expansion vessel is a sealed container in which a flexible diaphragm separates water from the cylinder and the air or gas which the sealed vessel contains. As the water in the cylinder heats,

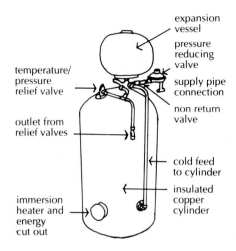

expansion vessel

pressure reducing valve

supply pipe connection

non return valve

temperature/ pressure relief valve

outlet from relief valves

cold feed to cylinder

insulated copper cylinder

immersion heater and energy cut out

Fig. 19 Unvented hot water storage cylinder.

Hot water supply pipework

HOT WATER SUPPLY SYSTEMS

it will expand against the diaphragm, compress the air and gas and so relieve the water expansion. In this way the expansion vessel acts in the same way as a vented expansion pipe to indirect systems.

The pressure relief, or reducing valve, is a safety device against the expansion vessel being unable to take up the whole of the expansion of heated water. For control against overheating there is a thermostat on the immersion heater and also a temperature limiting cut-out that operates on the electricity supply, and a temperature operated relief valve to discharge if the other controls fail.

Fig. 19 is an illustration of an unvented copper indirect cylinder prepared ready for installation as a packaged unit complete with factory applied insulation, expansion vessel and the necessary valves and controls.

The hot water supply pipework is run from the top of the cylinder with horizontal branches to the hot water outlets to fittings on each floor. Stop valves and draw-off taps are fitted to control and drain the pipe system as necessary for maintenance and repair.

There are two hot water supply systems, the central and the local. In the former, water is heated and stored centrally for general distribution, and in the latter water is heated, or heated and stored locally for local use. The difference between these systems is that with the central system hot water is run to the site of the sanitary appliances from a central heat source, and with the local system the heat source, gas or electricity, is run to the local heater which is adjacent to the sanitary appliances. Fig. 20 illustrates the two systems diagrammatically.

The central system is suited, for example, to houses, hotels, offices and flats where a central boiler fired by solid fuel, oil, gas or electricity heats water in bulk for distribution through a straightforward vertical distributing pipe system with short draw-off branches leading to taps to sanitary appliances on each floor. In large buildings, one heat source may serve two or more hot water storage cylinders to avoid excessively long distribution pipe runs.

The local system is used for local washing facilities where the fuel – gas or electricity – is run to the local heater either to avoid extensive and therefore uneconomic supply or distributing pipe runs, or because local control is an advantage.

In some buildings it may be economic to use a combination of central and local hot-water systems.

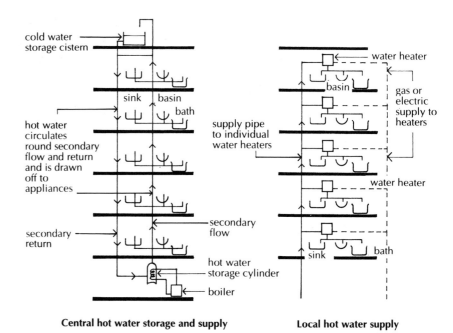

Fig. 20 Hot water supply systems.

Central hot water storage and supply   **Local hot water supply**

Central hot water supply

From Fig. 20 it will be seen that water is heated and stored in a central cylinder from which a pump circulates it around a distributing pipe system from which hot water is drawn.

In the two-storey house used to illustrate hot water distribution or supply pipe systems, the hot water was drawn directly from single branches. In a small building, such as a house, where the sanitary fittings are compactly sited close to the cylinder, the slight inconvenience of running off the cooled water in the single branches before hot water is discharged is acceptable.

In larger buildings such as that illustrated in Fig. 20, the inconvenience of running off cooled water from long single branches is unacceptable as it is also wasteful of both water and energy.

Dead leg draw-off

Because the water in single pipe branches to draw-off taps cools, these branches are termed dead leg branches or pipes, or dead leg of pipe. To avoid too great an inconvenience and waste of water and energy, the length of these dead legs is limited to 20 m for 12 mm pipes ranging to 3 m for pipes more than 28 mm, unless the pipes are adequately insulated against loss of heat.

The storage cylinder contains hot water sufficient for both anticipated peak demand and demands during the recharge period. The system is therefore designed to supply hot water on demand at all times. The one disadvantage of the system is that there is some loss of heat from the distributing pipes no matter how adequately they are insulated. This is outweighed by the economy and convenience of one central heat source that can be fired by the cheapest fuel available, and one hot water source to install, supply and maintain – hot water being at hand constantly by simply turning a tap.

Where a mains pressure supply system is used the supply pipe connects to the unvented hot water storage cylinder from which supply pipes connect to the fittings and there is no roof level storage system.

Local hot water supply

A water heater, adjacent to the fittings to be supplied, is fired by gas or electricity run to the site of the heater. Fuel for local heaters is generally confined to gas or electricity. The water is either heated and stored locally or heated instantaneously as it flows through the heater. The advantages of this system are that there is a minimum of distributing pipework, initial outlay is comparatively low and the control and payment for fuel can be local, an advantage, for example, to the landlord of residential flats. The disadvantage is that local heaters are appreciably more expensive to run and maintain than one central system.

Hot water storage heaters

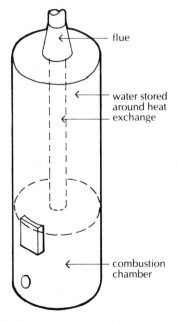

There are two types of local water heater, the hot water storage heater and the instantaneous water heater. The local hot water storage heater consists of a heat source and a storage cylinder or tank, and the instantaneous heater of a heat source through or around which cold water runs and is heated instantaneously as it is run off. The larger water storage heaters are used to supply hot water to ranges of fittings such as basins, showers and baths used in communal changing rooms of sports pavilions and washrooms of students' hostels.

The large, gas-fired, water storage heater illustrated in Fig. 21 consists of a water storage cylinder through which a heat exchanger rises to a flue from a combustion chamber. A thermostat in the water storage chamber controls the operation of the gas burners in the combustion chamber, cutting in to fire when the temperature of the water falls. A cold water supply pipe is connected to the base of the storage cylinder.

Hot water is drawn off either through a dead-leg draw-off pipe where pipe runs are short, or by a circulating secondary pipe system where runs are lengthy. The storage heater is heavily insulated to conserve energy. The size of the heater is determined by the anticipated use of hot water at times of peak use.

flue

water stored around heat exchange

combustion chamber

Fig. 21 Gas water storage heater.

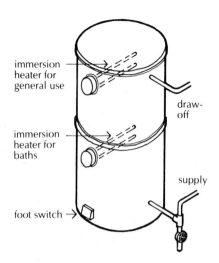

Fig. 22 Electric water storage heater.

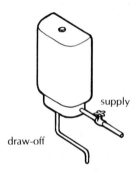

Fig. 23 Electric storage single point water heater.

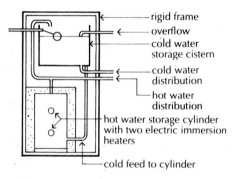

Fig. 24 Combined cold and hot water storage unit.

Instantaneous water heaters

The electric water storage heater, illustrated in Fig. 22, is for use in communal washrooms of students' hostels and residential schools where peak demand for hot water in bulk is generally confined to mornings and evenings, between which times the heater automatically reheats the water. These heaters, which are heavily insulated to conserve energy, are housed in a separate enclosure away from the wet activities they serve, for safety reasons. Hot water for basins is drawn from the top of the cylinder, which is heated by an upper immersion heater. Hot water in bulk for baths is boosted by the operation of both the upper and lower immersion heaters. The lower immersion heater is operated by a foot switch, a thermostat or a timed switch.

An advantage of the electric storage heater is that it does not have to be fixed close to an outside wall, which the gas heater does because of its flue.

The small electric, single point water heater, illustrated in Fig. 23, is designed to heat and store a small volume of water for the supply to single basins. The hot water storage cylinder and electric immersion heater, which are heavily insulated to conserve energy, are housed in a glazed enamel metal casing for appearance sake. A thermostat controls the electric supply to the immersion heater, cutting-in to reheat the water as it is drawn off. These heaters are used for basins in single toilets where it is convenient to run water and an electrical supply for the occasional use of hot water.

A combined cold cistern and hot water storage heater that is designed to fit into a confined space is illustrated in Fig. 24. Inside a rigid frame a cold water storage cistern and an insulated hot water storage cylinder are combined with connections for water and electric supplies, and hot and cold water draw-off connections. These units are designed specifically for use in small flats where space is limited, and they can be fitted close to bathroom and kitchen.

A disadvantage of these units is that there is poor discharge of water from outlets because of the small head pressure of water from the cold water storage cistern. Because of the small capacity of the cold water cistern the frequent refilling of the cistern may be somewhat noisy in a confined space.

These water heaters operate by running cold water around a heat exchanger so that water is heated as it flows. The heat exchanger only operates when water is flowing, hence the name instantaneous water heater.

Because the temperature of the water at the outlet is dependent on the rate of flow of water there is a limitation on the rate of flow from the outlet if the water is to be hot. Consequently the rate of flow from these heaters is limited.

Instantaneous gas water heater

Fig. 25 Gas instantaneous single point water heater.

Most instantaneous water heaters are fired by gas, which is ignited by a pilot light immediately water flows, to provide hot water instantaneously. Cold water running through a coil of pipework, wrapped around a combustion chamber and heat exchanger over a gas burner, is heated by the time it reaches the outlet. These heaters are controlled by the cold water supply valve; when the valve is opened the flow of water opens a gas valve to ignite the burners to heat water.

A single point, gas, instantaneous water heater is designed to supply hot water to single fittings such as a basin or sink. These heaters are usually fixed above the fitting to be supplied, and the hot water is delivered through a swivel outlet. A typical single point heater is illustrated in Fig. 25. Because of their small output and limited use the air intake from the room and the exhaust outlet to the room is acceptable. Where effective draught seals are fitted to windows to rooms in which these heaters are fired, there should be permanent ventilation to the open air.

Multi-point instantaneous gas water heater

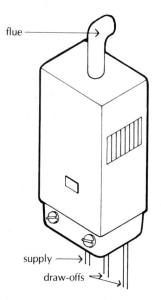

Fig. 26 Gas instantaneous multi-point water heater.

Multi-point gas water heaters were commonly used to supply domestic hot water to small houses and flats before combined space heating and hot water boilers became common. These heaters, illustrated in Fig. 26, can supply hot water to a sink, basin and bath through dead-leg draw-off pipes to fittings, hence the name multi-point.

When a tap over one of the fittings is opened, the flow of cold water through the coils of pipe around the heat exchanger is heated to deliver hot water. The initial flow of cold water opens a gas valve and the pilot light ignites the gas burners to provide heat. When all the taps to fittings are shut and there is no water flow, the gas valve shuts.

The rate of flow of hot water from these heaters is limited by the need for sufficient time to allow an adequate exchange of heat to the water coils in the heat exchanger. When more than one tap is opened there will be a restricted rate of flow of hot water, and filling a bath can be a somewhat lengthy process. For these reasons these heaters are much less used than they used to be.

The comparatively large output from these heaters necessitates a flue to open air to exhaust combustion gases and also an adequate intake of air for efficient and safe combustion of gases. A flue and permanent air vents or a balanced flue are necessary.

The gas valves in these heaters will only operate when there is comparatively high water pressure, such as that from a main supply, or a good head of water from a cistern. All gas fired heaters and boilers require regular maintenance if they are to operate efficiently and safely.

Instantaneous electric water heater

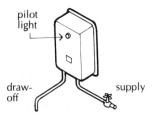

Fig. 27 Electric instantaneous single point water heater.

Water is heated as it flows through coiled heating elements immersed in a compact, sealed tank. The electric supply is operated by a flow switch in the cold water supply inlet. The control valve on the cold water supply pipe is used to adjust the temperature of the hot water outlet.

These heaters, illustrated in Fig. 27, are fixed over the single appliance to be supplied, with hot water delivered through a swivel outlet discharging over the appliance. The output of hot water from these heaters is limited by the rate of exchange of heat from the heating elements to cold water. In hard water areas limescale will coat the heating element and appreciably reduce the efficiency of these heaters. The heat exchange tank, which is heavily insulated, is housed in a glazed enamel casing for appearance sake.

The advantage of these heaters is that they are compact, require only one visible supply pipe and may be fixed in internal, unventilated toilets as they have no need for air intake or a flue.

WATER SERVICES TO MULTI-STOREY BUILDINGS

Mains water is supplied under pressure from the head of water from a reservoir, or pumped head of water or a combination of both. The level to which mains water will rise in a building depends on the level of the building relative to that of the reservoir from which the mains water is drawn, or relative to the artificial head of water created by pumps. Obviously mains pressure will rise less in a building on high ground than in one on lower ground.

In built-up areas there will, at times such as early morning, be a peak demand on the mains supply, resulting in reduced pressure available from the water main. It is the pressure available at peak demand times that will determine whether or not mains supply pressure is sufficient to feed cold water to upper water outlets. The pressure available varies from place to place depending on natural or artificial water pressure available, intensity of demand on the main at peak demand time and the relative level of the building to the supply pressure available.

Mains pressure supply

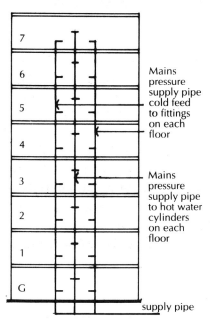

Mains pressure supply pipe cold feed to fittings on each floor

Mains pressure supply pipe to hot water cylinders on each floor

supply pipe

Fig. 28 Mains pressure supply.

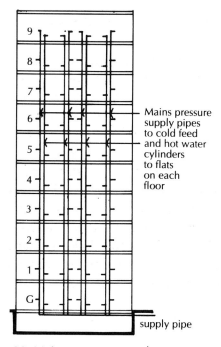

Mains pressure supply pipes to cold feed and hot water cylinders to flats on each floor

supply pipe

Fig. 29 Mains pressure supply.

Fig. 28 is a diagram of an eight-storey building where the mains pressure at peak demand times is sufficient to supply all cold and hot water outlets to all floors. Two supply pipe risers branch to provide cold water to ranges of sanitary fittings in male and female toilets on each floor and another riser branches to feed hot water storage cylinders on each floor for the toilets. This is the most economic arrangement of pipework where sanitary fittings are grouped on each floor, one above the other.

To provide reasonable equality of flow from outlets on each floor it is usual to reduce pipe sizes. In the arrangement shown in Fig. 28, the bore of the risers will be gradually reduced down the height of the building to provide a reasonable flow from all outlets to compensate for reduced flows as pumped head pressure increases down the height of the building.

Where the mains pressure is sufficient to provide a supply to a multi-storey building it is not always possible to provide a reasonable equality of flow to fittings on each floor by varying pipe sizes alone. An increase in pipe size will provide a little reduction in pressure loss from the frictional resistance to flow of larger bore pipes and fittings. There is a limit to the resistance to flow that can be effected, because of the limited range of pipe sizes, without using gross and uneconomic pipe sizes.

In the diagrammatic pipe layout illustrated in Fig. 29, of sanitary fittings in a ten-storey building with four flats to each floor, a pair of risers supplies the cold water outlets and hot water cylinder to each flat on each floor, with some reduction in pipe size down the building to provide reasonable equalisation of flow. Had a pair of risers been used to each pair of flats on each floor, it would not have been practical to provide reasonable equalisation of flow.

Another method of providing equalisation of rate of flow from outlets floor by floor in multi-storey buildings is by the use of pressure reducing valves at each flow level. This more sophisticated approach is used in modern buildings in Northern European countries.

Multiple risers

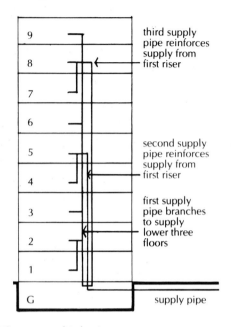

Fig. 30 Multiple risers.

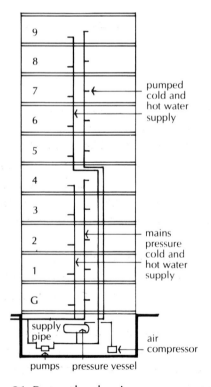

Fig. 31 Pumped and mains pressure supply.

Another method of equalising flow, being used experimentally, is the use of multiple rising pipes as illustrated in Fig. 30. One rising supply pipe branches to supply the three lower floors and then rises to supply the three floors above, where it is joined and its supply reinforced by the second rising supply and then up to the top three floors where it is joined and reinforced by the third rising supply pipe.

The logic in the use of this arrangement is that in multi-storey buildings in multiple occupation, such as a hotel, there will be unpredictable short periods of peak use, as for example when a tour party returns. During this short period there may be heavy call on water use on one floor, which may cause unacceptable starvation of water supply. By providing alternative reinforcing sources of supply and judicious arrangements of pipe sizes, a reduction of flow rate may be avoided or at least smoothed out.

It is good practice in the design of pipework layout to make an assumption of frequency of use of draw-off water to sanitary appliances. The assumption is based on an estimate of peak period use, such as first thing each morning, which does not allow for unpredictable heavy use. To provide for possible maximum use would involve gross and uneconomic pipework.

Where mains pressure is insufficient at peak demand times to raise water to the highest water outlet, it is necessary to install pumps in the building to raise water to the higher water outlets. In this situation it is usual to supply from the supply pipe those outlets that the mains pressure will reach, and those above by the pumped supply, to limit the load on the pumps as illustrated in Fig. 31.

Two mains pressure supply pipes rise to supply the lower five floors, one with branches to each floor level to cold water outlets and the other with branches at each floor level to hot water storage cylinders. The five upper floors are supplied by two rising supply pipes under the pressure of the pump in the basement. At each floor level branches supply cold water to sanitary fittings and there are branches to hot water cylinders.

There are two pumps, one operating and the other as standby in case of failure and to operate during maintenance. The pumps are supplied by the mains through a double check valve assembly to prevent contamination of the supply by backflow should the pumps fail.

Auto-pneumatic pressure vessel

The auto-pneumatic pressure vessel indicated in Figs. 31 and 32 is a sealed cylinder in which air in the upper part of the cylinder is under pressure from the water pumped into the lower part of the cylinder. The cushion of air under pressure serves to force water up the supply pipe to feed upper-level outlets as illustrated. Water is drawn from the auto-pneumatic pressure vessels as water is drawn from the upper-level outlets so that when the water level in the pressure vessel falls to a predetermined level, the float switch operates the pump to recharge the pressure vessel with water. Thus the cushion of air in the pressure vessel and its float switch control and limit the number of pump operations. In time, air inside the pressure vessel becomes mixed with water and is replaced automatically by the air compressor.

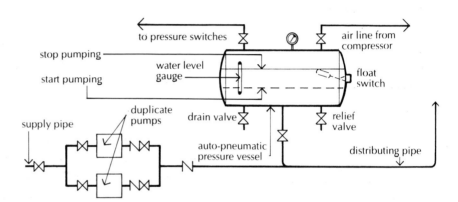

Fig. 32 Auto-pneumatic pressure vessel.

Gravity feed cold water supply

The traditional system of cold water supply was through a roof level cold water storage system to supply all cold water outlets except one mains pressure outlet to sinks for drinking water. The justification for this arrangement was that the air gaps between the mains supply and the water level in the system served as an effective and trouble free check against possible contamination of the mains supply. A separate drinking water supply was provided on the grounds that cistern water might not be palatable.

In multi-storey buildings, where mains pressure is insufficient to raise water to roof level, the comparatively trouble free, cistern feed to cold water outlets may be used with a covered drinking water storage vessel or cistern.

Fig. 33 is a diagram of a ten-storey building in which the drinking water outlets to the lower five floors are fed by mains supply. Because the mains pressure is insufficient to raise water to roof level the upper five floors are supplied with drinking water from a covered drinking water storage vessel or cistern. The roof level cold water storage cistern, the drinking water vessel or cistern and the drinking water outlets to the top five floors are fed by a pumped

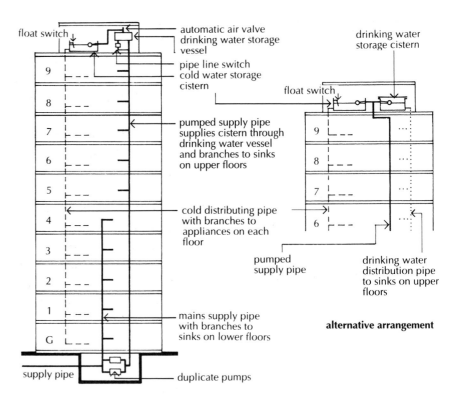

Fig. 33 Drinking water storage cistern.

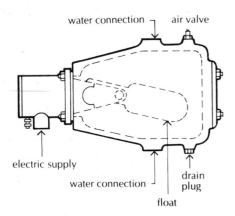

Fig. 34 Pipeline switch.

Drinking water storage cistern

supply. The cold water storage cistern supplies cold water to all cold water outlets, other than the sink and to hot water cylinders on each floor.

There is a duplication of rising pipework to the five lower floors, which is considered a worthwhile outlay in reducing the load and wear on the pumps. The pumped supply feeds both the higher drinking water outlets and cold water storage cistern through a drinking water storage vessel.

The sealed drinking water storage vessel and the cold water storage cistern are filled through the pumped supply, in which a pipeline switch is fitted. A pipeline switch, illustrated in Fig. 34, is used to limit pump operations. As water is drawn from the drinking water vessel the water level falls until the float in the pipeline switch falls and starts the pump.

The cold water storage cistern is supplied through the drinking water vessel, from which it can draw water to limit pump operations. When the water level in the cold water storage cistern falls to a pre-determined level, a float switch starts the pump to refill the cistern through the drinking water vessel.

As an alternative a sealed drinking water storage vessel – a drinking water storage cistern – may be used, as illustrated in Fig. 33. The cistern has a sealed cover and filtered air vent and overflow to

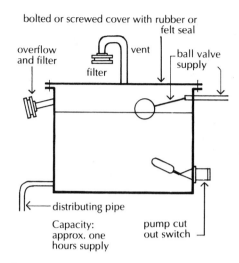

Fig. 35 Drinking water storage cistern.

exclude dirt and dust, as illustrated in Fig. 35. A pump cut out, or float switch, controls the pump operations at a predetermined level. As before, a float switch controls pump operations to fill the cold water storage cistern. The pumped service pipe feeds both cisterns so that whichever switch operates, the pump operates to fill both cisterns.

The advantage of the sealed drinking-water storage vessel is that it requires less maintenance than the drinking water cistern whose ball valve and filters require periodic maintenance. But the roof level switches have to be wired through a control box down to the basement level pumps and the switches will require regular maintenance in a position difficult to access.

As a check to the possibility of backflow from a pumped supply into the main, and as a reservoir against interruption of the mains supply, it has been practice to use a low level cistern as feed to the pumped supply to a roof level cistern, so that the air gap between the inlet and the water level in the cistern acts as a check to possible contamination of the mains supply.

Low level cistern

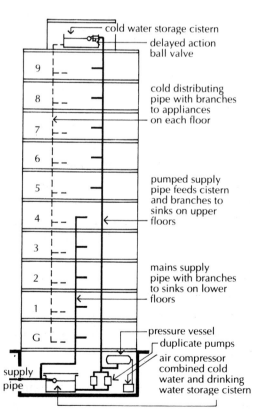

Fig. 36 Low level storage cistern.

Where the mains pressure is insufficient to supply cold water outlets on upper floors of multi-storey buildings, and a separate drinking water supply is required, a cistern feed supply to cold water outlets may be used in combination with a pumped supply to upper level drinking water outlets, as illustrated in Fig. 36.

Here a low level cistern is fitted as supply to the pumps. A low level cistern is used as a form of check against contamination of the supply and as a standby against interruption of the mains supply. The covered low level cistern serves to supply the upper level drinking water outlets and the roof level storage cistern. The operation of the pumps is controlled at low level through a pressure vessel similar to that illustrated in Fig. 32.

Drinking water outlets to the lower floors are connected to the mains supply pipe which in turn feeds a low level storage cistern from which a supply is pumped to the upper level drinking water outlets and the roof level storage cistern. Pump operations are limited by an auto-pneumatic pressure vessel, illustrated in Fig. 32, and a delayed action ball valve to the roof level cistern. As the low level cistern supplies both drinking and cold water outlets, it has to be sealed to maintain the purity of the drinking supply. A screened air inlet maintains the cistern at atmospheric pressure.

The pumped supply pipe shown in Fig. 36 feeds the roof-level cold water cistern, which is fitted with a delayed action ball valve.

Delayed action ball valve

A delayed action ball valve, illustrated in Fig. 37, is fitted to the roof level cistern to control and reduce pump operations. The delayed action ball valve consists of a metal cylinder (A) which fills with water when the cistern is full and in which a ball (B) floats to operate the valve (C) to shut off the supply. Water is drawn from the cistern and as the water level falls, float (E) falls and opens valve (D) to discharge the water from cylinder (A). The ball (B) falls and opens the valve (C) to refill the cistern through the pump, so limiting the number of pump operations.

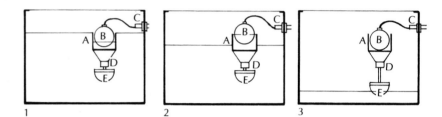

Fig. 37 Delayed action ball valve.

Intermediate water storage cistern

Current practice in multi-storey buildings of more than about ten storeys is to use a roof level and one or more intermediate water storage cisterns to supply outlets other than drinking water taps. The intermediate cisterns spread the very considerable load of water storage and also serve to reduce the pressure in distributing pipes, for which reason they are sometimes termed 'break pressure cisterns'.

Fig. 38 shows a ground-level storage cistern supplying a pumped supply to an intermediate and roof-level cistern from which distributing pipes supply sanitary appliances on the lower and upper floors respectively, thus spreading the weight of water storage and limiting pressure in distributing pipes. The pumped supply also feeds drinking water outlets to upper floors. A float switch in the pressure vessel and delayed-action ball valves in the cisterns limit pump operations. Intermediate cisterns are used at about every tenth floor.

Zoned supply system

The supply to the 22 storey building illustrated in Fig. 39 is divided into three zones in order to help equalise pressure and uniformity of flow from outlets on all floors. The supply to the lower nine floors is through two rising supply pipes taken directly from the mains supply.

The supply to the eight top floors is from a pumped supply through a roof level pump and pressure vessel, and the supply to the intermediate five floors is by gravity from a sealed drinking water cistern at roof level. In this way the loss of residual head to the lower

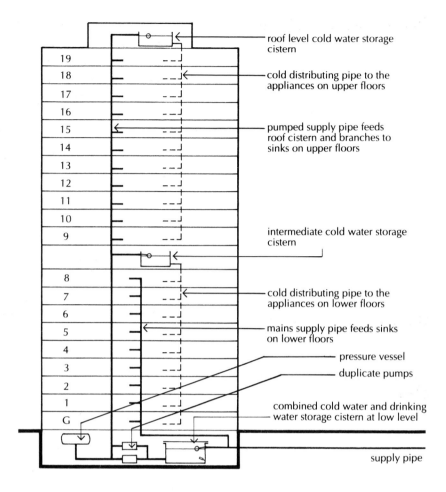

Fig. 38 Intermediate level storage cistern.

and upper floors is limited and the loss of head from the cistern is limited by feeding intermediate floors only. By dividing the building into three zones of supply, loss of head is limited in the main supply and distribution pipes, and by reduction in pipe diameter in each zone reasonable equalisation of flow from taps on each floor is possible.

PIPES (TUBULARS) FOR WATER SUPPLY

The materials used for pipework for water supplies are copper, galvanised mild steel and plastic. Up to the middle of this century lead was the material most used for water-supply pipework. The increase in the cost of lead after World War II led first to the use of galvanised mild steel tubes and later to light-gauge copper tubes instead of lead. Today, the light gauge copper tube is the material most used for water-supply pipework in buildings.

Minute quantities of lead may be leached from lead pipes used for water supplies. These small amounts of lead may in time be sufficiently toxic, particularly to the young, to be a serious hazard to health. The new Water Supply Byelaws prohibit the use of lead for

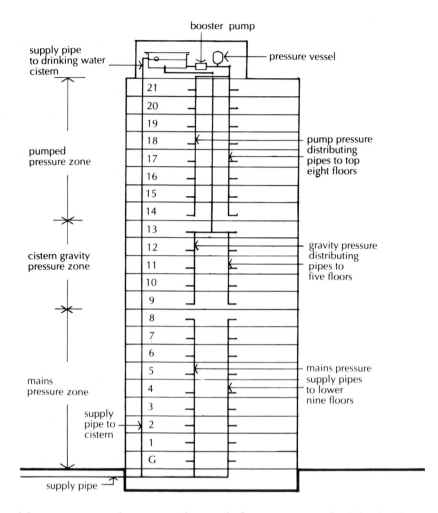

Fig. 39 Zoned supply system.

either new or replacement pipework for water supply. The byelaws also prohibit the use of lead solders for joining copper pipes, and so tin/silver solders are accepted.

Copper pipe (copper tubulars)

The comparatively high strength of copper facilitates the use of thin-walled light-gauge pipes or tubes for most hot and cold water services. The ductility of copper facilitates cold bending, the thin walls make for a lightweight material and the smooth surface of the pipe provides low resistance to the flow of water. Like lead, copper oxidises on exposure to air and the oxide film prevents further corrosion. Copper tube is the material most used for water services today.

Copper tubes are supplied as half-hard temper and dead soft temper, the former for use above ground for its rigidity and the latter principally for use in trenches or where its flexibility is an advantage, as in use in the underground service to buildings. Light gauge copper tube, size 6 to 159 mm (outside diameter) is manufactured; the sizes most used in buildings are 12, 15, 18, 22, 28, 35, 42 and 54 mm.

Jointing

Copper pipes are joined with capillary or compression joints – used for the majority of copper tubes in water services – and welding, for the larger bore pipe for drains above ground where the pipework is prefabricated in the plumbers' shop.

Capillary joint

A capillary joint is made by fitting plain ends of pipe into a shouldered brass socket. Molten solder is then run into the joint, or internal solder is melted by application of heat. Pipe ends and fittings must be clean, otherwise the solder will not adhere firmly to the pipe and socket. Fig. 40 illustrates typical capillary joints. This is a compact, neat joint.

Capillary joints, which afford the least labour and cost of material, are used for most small bore pipework particularly where there are a lot of joints.

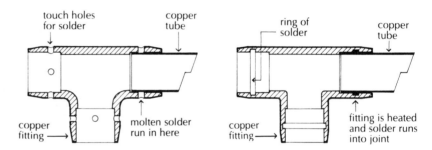

Fig. 40 Soft solder capillary fittings.

Compression joint

Compression fittings are either non-manipulative or manipulative. With the former, plain pipe ends are gripped by pressure from shaped copper rings; in the latter the ends of the pipes are shaped to the fitting, as illustrated in Fig. 41.

The manipulative fitting is used for long pipe runs and where pipework is not readily accessible, as the shaped pipe ends are more firmly secured than with the non-manipulative or capillary joint. The somewhat more expensive non-capillary joint is sometimes preferred to the cheaper capillary joint which is dependent on cleanliness of contact rather than friction.

Compression joints are often combined with capillary joints at the ends of pipe runs to facilitate repairs or alterations, because of the ease of disconnecting these joints.

The thin walled copper tubulars used for water services have poor mechanical strength in supporting the weight of a filled pipe. The pipework should, therefore, be supported at comparatively close

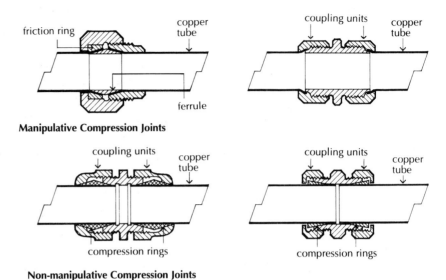

Manipulative Compression Joints

Non-manipulative Compression Joints

Fig. 41 Compression joints.

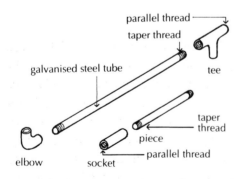

Fig. 42 Copper fixing clips.

intervals by pipe clips screwed to plugs in walls. The pipe clips, illustrated in Fig. 42, should be fixed at intervals of from 1.5 to 3 m horizontally and 2 to 3 m vertically, depending on the size of the pipe.

Galvanised mild steel (low carbon steel)

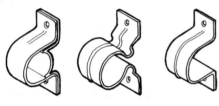

Fig. 43 Galvanised steel pipe.

Jointing

Galvanised mild steel (low carbon steel) tubulars were commonly used for all water services during the middle of this century because of the low cost and availability of the material. Today, galvanised mild steel tubulars are used for service, supply and distributing pipework, 40 mm bore and over, particularly where there are long straight runs of pipework, because of the economy of material and labour in the use of this material. Galvanised mild steel tubulars are also used for fire prevention and fire fighting installations because of their greater resistance to damage by heat than other pipe materials.

Mild steel tubulars are manufactured with a nominal bore of 6 to 150 mm in three grades: light, banded brown; medium, banded blue; and heavy, banded red. In general, light is for gas services, medium for water services, and heavy for steam pressure services. For water services, tubulars should be galvanised to resist corrosion. Fig. 43 shows these tubulars.

The pipe ends are threaded and the joint made with sockets, nipples or long screws, unions or fittings as illustrated in Fig. 43. Pipes are supported at intervals of from 2.5 to 3 m.

Plastic pipe or tube

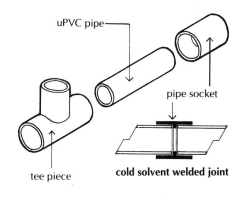

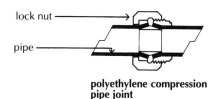

Fig. 44 PVC pipe.

Polythene and unplasticised PVC (poly vinyl chloride) tube is used respectively for use underground and above ground for water services.

Polythene (polyethylene) is flexible and used for water services underground, while the more rigid PVC is used for services above ground. Both materials are lightweight, cheap, tough, do not corrode and are easily joined. These materials soften at comparatively low temperatures and are used principally for cold water services and drains above and below ground. The pipes are manufactured in sizes from 17 to 609 mm, the sizes most used being 21.2, 26.6, 33.4, 42.1 and 60.2 mm (outside diameter).

uPVC and ABS (acrylonitrile butadine styrene) pipes are used primarily for waste pipes.

Polythene tube is jointed with gunmetal fittings as illustrated in Fig. 44. A copper compression ring is fitted into the pipe ends to prevent the wall of the pipe ends collapsing as the coupling is connected by tightening the nut. This type of joint is used to withstand high pressures, as in mains pressure supply pipework.

Solvent welded joint

A solvent weld cement is applied to the ends of the pipes to be joined and the ends of the pipes are fitted into the socket or other fitting (Fig. 44). The solvent cement dissolves the surfaces of the pipe end and fitting so that as the solvent hardens it welds the joined plastic surfaces together. The joint will set in 5 to 10 minutes and require 12 to 24 hours to become fully hardened.

This type of joint is commonly used with uPVC pipework, such as low pressure distribution pipework and more particularly waste branch pipes. Pipe runs in plastic are supported by clips at 225 to 500 mm horizontally and 350 to 900 mm vertically.

Insulation

Cold and hot water supply and distributing pipes should be fixed inside buildings, preferably away from external walls to avoid the possibility of water freezing, expanding and rupturing pipes and joints. Where pipes are run inside rooftop tank rooms and in unheated roof spaces, they should be insulated with one of the wrap-around or sectional insulation materials designed for the purpose. Similarly, roof-level and roof-space storage cisterns should be fitted with an insulation lining to all sides, the base, and to the top of the cisterns.

VALVES AND TAPS

Valves

The two general terms used to describe fittings designed to regulate or shut the flow of water along a pipeline, are valve and cock. A valve is a fitting that can be adjusted either to cause a gradual restriction in flow or a cessation of flow. For this reason valves are also known as stop valves.

Cock, plug cock, stop cock

A cock is a fitting that consists of a solid cylindrical plug, fixed with its long axis at right angles to flow. When shut the plug fills the pipe line. A 90° quarter turn of the plug brings a round port or hole in the plug into line with the flow, and so opens the plug. There is little control of flow with a plug that is designed to be either shut or open. Cocks are also described as plug cocks and stop cocks.

Globe valves

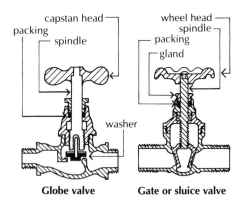

Fig. 45 Valves.

The two valves in common use are the globe valve and the gate valve.

Globe valves, so named for their rounded or globe shape, operate to control or shut flow, through a disc that is lowered slowly by turning a screwdown spindle to a seating. Because the operation to shut these valves is by screwing down, they are also described as screw down valves.

The design of these valves is such that they operate to close and open at right angles to the line of flow. In consequence there is a little restriction of flow when the valve is open as water flows up and down and then along the pipe line.

A globe valve is illustrated in Fig. 45. Globe valves are commonly used in high pressure and hot water pipework.

Gate valve, fullway gatevalve

A gate valve operates by raising or lowering a metal gate into or out of the line of the pipework as the spindle is screwed down or up. This valve is sometimes referred to as a fullway gate valve as when it is fully open it does not restrict flow along the pipeline, unlike the globe valve. For this reason the gate valve is used where there is low pressure flow in the pipeline, such as that from cistern feed systems. Fig. 45 is an illustration of a gate valve.

Taps

The word tap is used in the sense of drawing from or tapping into, as these fittings are designed to draw hot or cold water from the pipework. They are sometimes described as draw-off taps.

There are three types of draw-off tap: the bib tap, the pillar tap (including mixer taps which mix hot and cold water) and the draw-off tap which is used to draw off water to drain pipe systems.

Bib tap

The traditional bib tap, illustrated in Fig. 46, operates in the same way as a globe valve through a disc which is screwed down to close and up to open, except that the tap is at the end of a pipeline to open to discharge water. A washer fixed to the base of the disc seals the tap when shut. The supply to the bib tap is in line with the pipeline.

This type of tap is fixed above the bath, basin or sink to be served.

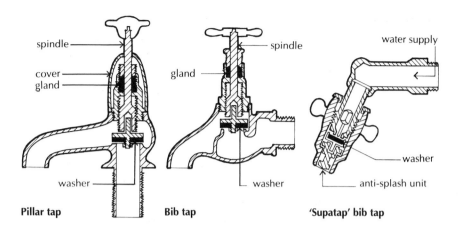

Fig. 46 Taps. **Pillar tap** **Bib tap** **'Supatap' bib tap**

Pillar tap

This tap, illustrated in Fig. 46, is designed to be fitted to the bath, basin or sink it serves, with the supply pipe connected vertically to the base of the tap. This type of tap is usually covered by a metal or plastic cover for the sake of appearance. A wide variety of designs is available. The tap operates in the same way as a bib tap, through a screw down spindle which opens or closes a washer.

Supatap

The 'supatap', illustrated in Fig. 46, is designed so that the washer may be replaced without turning off the water supply, through an automatic closing valve that interrupts the supply.

Quarter turn taps

Many modern taps use a system of ceramic discs to open and close the supply. The lower disc is fixed and the top disc can be turned through 90°. When the two ports or holes in the top disc coincide with those in the bottom fixed disc, the water flows. A lever, when turned or depressed, operates a spindle to effect the necessary quarter turn to open the tap for water to flow.

Because of the quarter turn operation there is not the same control of flow as there is with bib and pillar taps and because of the small ports through which water flows there is not the same vigorous flow.

The advantage of these taps is that the polished ceramic discs will last as long as the tap itself without need of replacement. The simple operation of these taps makes it possible to design plain, elegant taps that consist of a plain steel cylinder on which a lever operates to open and shut the tap and can be used to operate a hot and cold water mixer.

Fig. 47 is an illustration of a quarter turn tap with ceramic disc washers.

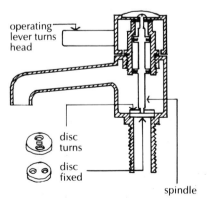

Fig. 47 Quarter turn tap.

Draw-off taps

A draw-off tap is similar to a bib tap except that it is operated by a loose wheel or capstan head and the outlet discharge is serrated to take a hose connection.

Contamination of supply

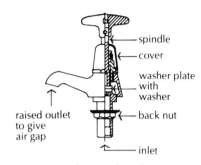

spindle
cover
washer plate with washer
raised outlet to give air gap
back nut
inlet

Fig. 48 Tap with raised outlet.

As one of several requirements to prevent contamination of the supply by backflow, the Water Supply Byelaws require an air gap between the lowest part of the outlet of water taps and the spill over level of sanitary appliances such as basins, baths and sinks. The rationale for this regulation is somewhat difficult to understand. It seems highly unlikely that a wash basin overflow, and overflow water rising above the rim of the basin, would coincide with the reduction or cessation of the supply and so cause contaminating backflow. Fig. 48 is an illustration of a tap to match this requirement.

The size of the gap is related to the bore of the pipe supplying the tap or the bore of the outlet. As an alternative to an air gap, a double check valve assembly or other suitable appliance may be used in the pipe supplying each tap.

ESTIMATION OF PIPE SIZES

Water storage capacity

British Standard 6700, 1987, recommends minimum storage for dwellings of 100–150 litres for small houses, for cold water only, and 100 litres per bedroom, total capacity, for larger houses.

Water storage capacity was determined by the regulations of water undertakings for dwellings at 114 and 227 litres for cold and domestic hot water respectively and for other uses from Table 2.

To provide a reasonable rate of flow from outlets, that is taps and valves, the required size of pipe depends on the static or pumped head of water pressure, the resistance to flow of the pipes, fittings and bends and the assumed frequency of use of outlets. For small pipe installations such as the average dwelling where there are 5 to 10 outlets, pipes of sufficient bore are used to allow simultaneous use of all outlets at peak use times. As only small bore pipes are required for this maximum rate of flow there is no point in making an estimate of pipe size for a more realistic usage.

With larger installations, such as the pipe system for a block of flats where there may be 100 or more outlets, it would be unrealistic and uneconomic in pipe size and cost to assume that all outlets will be in use simultaneously. It is usual therefore to make an assumption of the frequency of use of outlets, to estimate required pipe sizes that will give a reasonable rate of flow from outlets that it is assumed will be in use simultaneously at peak use times. If the actual simultaneous use is greater than the estimate then there will be reduced rate of flow from outlets. This 'failure' of the pipe system to meet actual in-use flow rates has to be accepted in any estimate of frequency of use of outlets.

Table 2 Cold water storage.

Recommended minimum storage of cold water for domestic purposes (hot and cold outlets)	
Type of building or occupation	**Minimum storage litres**
Hostel	90 per bedspace
Hotel	200 per bedspace
Office premises with canteen facilities without canteen facilities	45 per employee 40 per employee
Restaurant	7 per meal
Day school nursery primary	15 per pupil
secondary technical	20 per pupil
Boarding school	90 per pupil
Children's home or residential nursery	135 per bedspace
Nurses' home	120 per bedspace
Nursing or convalescent home	135 per bedspace

Rate of flow

The rate of water flow at taps and outlets depends on the diameter of the outlet and the pressure of water at the tap or outlet. The size of the tap is fixed. The water pressure depends on the source water pressure from a cold water storage cistern or a pumped supply, and the loss of pressure to the frictional resistance of the pipework and its fittings such as elbows, tees, valves and taps. The design of pipework installation is concerned, therefore, with estimating the resistance to flow and the selection of pipes of sufficient size to allow a reasonable rate of flow at taps, where the source water pressure is known.

Water pressure hydraulic or static head

In the design of pipework for buildings it is convenient to express water pressure as hydraulic or static head, which is proportional to pressure. The static head is the vertical distance in metres between the source, the cold water storage cistern and the tap or outlet. This head represents the pressure or energy available to provide a flow of water from outlets against the frictional resistance of the pipework and its fittings. The frictional resistance to flow of pipes is expressed as loss of head (pressure) for unit length of pipe.

Loss of head

These loss of head values are tabulated against the various pipe diameters available and the various material in use. The frictional

Table 3 Equivalent pipe lengths.

Pipe fittings	Equivalent length of pipe in pipe diameters
90° Elbows	30
Tees	40
Gate valves	20
Globe valves and taps	300

resistance to flow of fittings such as elbows, tees, valves and taps is large in comparison to their length in the pipe run. To simplify calculation it is usual to express the frictional resistance of fittings as a length of pipe whose resistance to flow is equivalent to that of the fitting. Thus the resistance to flow of a tee is given as an equivalent pipe length.

The equivalent length of pipe is given in Table 3. From this the frictional resistance of a pipe and its fittings can be expressed as an equivalent length of pipe, that is the actual length plus an equivalent length for the resistance of the fittings. The head (pressure) in that pipe can then be distributed along the equivalent length of pipe to give a permissible rate of loss of head per metre run of equivalent pipe length, and the head remaining at any point along the pipe can be determined. From this, the pipe diameter required for a given rate of flow in pipework and at outlets can be calculated.

To select the required pipe sizes in an installation it may be useful to prepare an orthographic or isometric diagram of the pipe runs from the scale drawings of the building. This diagram need not be to scale as the pipe lengths and head available will, in any event, be scaled off the drawings of the building. The purpose of the diagram is for clarity in selecting pipe sizes and tabulating these calculations.

Fig. 49 is an isometric diagram, not to scale, of a cold distribution pipe installation for a small building. The source of supply – the cold water storage cistern – is shown. The head from the base of the cistern is measured and all pipe runs to sanitary appliances are indicated. Each pipe run is numbered between tees and tees and tees and taps.

A change of pipe diameter is most likely to be required at tees and it is convenient, therefore, to number pipes between these points and taps. There are various methods of numbering pipe runs. The method adopted in Fig. 49 is a box, one corner of which points to the pipe or one side of which is along the pipe run. The box contains the pipe number on the left hand side, the actual pipe length top right and the rate of flow in litres per second bottom right.

The rate of flow in a pipe is the rate of flow of the single sanitary appliance it serves, or the accumulation of all the rates of flow of all the sanitary appliances it serves. In Fig. 49 the head is measured from the base of the cistern to the taps or pipe runs. Some engineers measure from midway between the water line and the base of the cistern and others from some short distance below the cistern to allow a safety margin and to allow for furring of pipes. For most building installations it is usual to take head measurements from the base of the cistern.

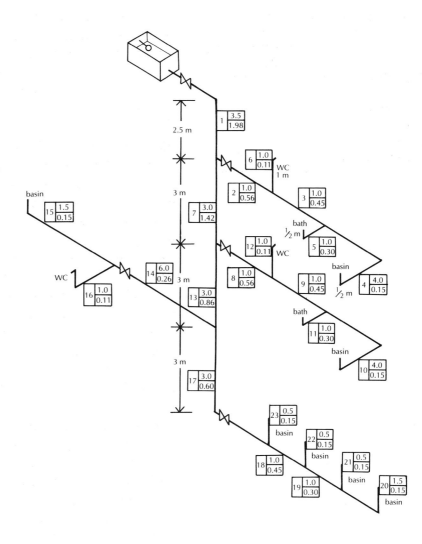

Fig. 49 Diagram of cold distribution pipe layout.

Frequency of use

In the following calculations to determine pipe sizes it is assumed that it is possible that all the taps served may be open simultaneously, and the pipes are sized accordingly. In large installations such as those of multi-storey blocks of flats, it is unlikely that all taps will be open at the same time and a frequency of use assumption, described later, is made to avoid the expense of over-large pipes.

Critical run of pipes

The procedure for selecting pipe sizes is firstly to determine the first index or critical run of pipes – that along which taps are most likely to be starved of water when all the taps are open. This starvation or loss of rate of flow is most likely to occur in the branch closest to the cistern, where head pressure is least. The first index run in Fig. 49 is pipe run 1, 2, 3, and 4. If pipes of adequate size for this run are selected to provide the required rates of flow at taps, then pipe run 1 will be large enough to supply all the other taps to the rest of the installation.

The head of water available in the fist index run 1, 2, 3, and 4 is 2 m, the vertical distance from the base of the cistern to the tap of the wash basin. The rate of flow from taps determines the cumulative rate of flow in the pipe runs. The unknown factors are the frictional resistance of the pipework and fittings and the size of pipes required for the given rates of flow. It is necessary, therefore, to make an initial assumption of one of these unknowns – the pipe sizes.

Assumption of pipe sizes

To make this initial assumption it is necessary to calculate a rate of loss of head in the index pipe run. The actual length of pipes is known and it is necessary to make an estimate of the likely length of pipe whose resistance to flow is equivalent to the resistance of the pipe fittings. This is usually taken as a percentage of the actual pipe length, and may vary from 25 to over 100. With experience in pipe sizing this assumed percentage will approach a fair degree of accuracy. In general, the greater the number of fittings to each unit length of pipe, the higher the percentage.

In Fig. 49, pipe run 1, 2, 3 and 4 has an actual length of 9.5 m. Assume an equivalent length of about 50% or say 5.5 m. Thus the total equivalent length of pipe is 9.5 + 5.5 = 15 m. The head is 2 m. The permissible rate of loss of head is therefore 2/15 = 0.13 per metre of equivalent length of pipe and this rate of loss of head should not be exceeded at any point along the pipe run.

The graph in Fig. 50 is used to select pipe sizes where the rate of flow is known and rate of loss of head has been assumed. In our example the rate of loss of head is 0.13. From that point on the base line read up to 1.98 litres per second, the flow required in pipe 1. These two intersect roughly midway between the heavy oblique lines indicating 35 and 42 mm pipe sizes. If the 35 mm pipe were selected then the rate of loss of head would be greater than the permissible loss of head figure of 0.13, so the next larger size, 42 mm, is selected for pipe run 1l.

Similarly, for pipe runs 2, 3 and 4, from the 0.13 rate of loss of head on the graph read up to 0.56, 0.45 and 0.15 rates of flow to select pipe sizes 28, 28 and 18 mm respectively, choosing as before the pipe size above the intersection of points.

These assumed pipe sizes may now be used to make a more exact calculation of friction losses in pipes and fittings and a more exact selection of pipe sizes. For this purpose it is usual to tabulate the calculations, as set out in Fig. 51.

The pipe numbers, design flow rates and actual pipe lengths of pipe runs 1, 2, 3 and 4 – the first index run – are entered in columns 1, 2, and 3 and the assumed diameters in column 7. The friction losses for

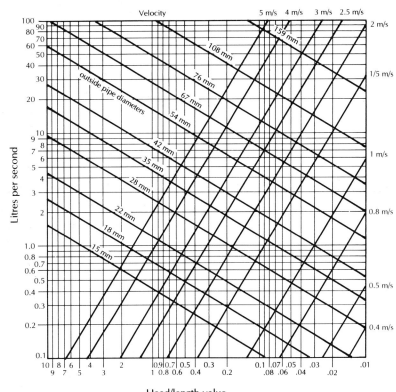

Velocity

Litres per second

Head/length value
Rate of loss of head per metre

Fig. 50 Graph for selection of copper pipe sizes. Pipe sizes are outside diameters.

1	2	3	4	5	6	7	8	9	10	11	12
Pipe No.	Design flow	Length	Equiv. length	Total eq. length	Head available	Assumed diam.	Permissible H/L value	Actual H/L value	Head used	Head remaining	Final diam.
1	1.98	3.5	3.8	7.3	2.0	42	0.09	0.085	0.62	1.38	42
2	0.56	1.0	1.7	2.7	1.38	28	0.09	0.06	0.16	1.22	28
3	0.45	1.0	1.1	2.1	1.22	28	0.09	0.04	0.08	1.14	28
4	0.15	4.0	6.5	10.5	1.14	18	0.09	0.06	0.63	0.51	18
5	0.30	1.0	6.0	7.0	1.14	18	0.16	0.08			22
6	0.11	1.0	5.0	6.0	0.72	15	0.12	0.09			15
7	1.42	3.0	1.4	4.4	4.88	35	1.1	0.35	1.54	3.34	28
8	0.56	1.0			3.34						
9	0.45										
10	0.15										
11											

Fig. 51 Table for calculation of pipe size.

each pipe run due to elbows, tees, valves and taps are now calculated. These may be determined by a multiplier of the assumed pipe diameters to give an equivalent length of pipe from Table 3, or more accurately from tables published by the Chartered Institution of Building Services Engineers.

In pipe run 1 there is a gate valve, an elbow and a tee; the equivalent pipe sizes for these, from Table 3, are 20, 30 and 40 respectively. The sum of these, 90, multiplied by the assumed pipe diameter of 42 gives an equivalent pipe length of 90 × 42 or 3.8 m. Thus the total actual length of pipe 1, and equivalent length for fittings, is 3.5 + 3.8 = 7.3 m, and this figure is entered in column 5. Similarly, for pipe run 2 where there is a valve and a tee, 20 + 40 = 60, which multiplied by the assumed pipe diameter of 28 gives an equivalent length of 1.68 m, say 1.7 m, which is entered in column 4 to give a total length of 2.7 m in column 5. For pipe run 3 there is one tee, 40, which gives an equivalent length of 40 × 28 = 1.12 m, say 1.1 m, to give a total equivalent length of 2.1 m. In pipe run 4 there are two elbows, 30 + 30, and one tap, 300, a total of 360 which multiplied by the assumed pipe diameter of 18 equals 6.48 m, say 6.5 m, to give a total equivalent length of 10.5 m.

With these more accurate totals of equivalent lengths in column 5 it is now possible to make a more accurate calculation of rate of loss of head. The head is 2 m and the total of equivalent lengths for pipe runs 1 to 4 in column 5 is 22.6. Thus the permissible rate of loss of head is 2 ÷ 22.6 = 0.09, which is entered in column 8. With this more accurate rate of loss of head value, final pipe sizes are selected from Fig. 51.

For pipe run 1 the head loss and flow lines intersect just below the 42 mm pipe size line, confirming the assumed pipe size. The 42 mm pipe size line intersects the 1.98 flow line above a head loss value of 0.085 so that the actual head loss value in selecting the 42 mm pipe will be less than the permissible head loss value. The actual head loss in pipe run 1 will, therefore, be 0.085, which multiplied by the total equivalent length of 7.3 is 0.62, and this figure is entered in column 10 to give a figure of 1.38 head remaining in column 11.

This figure of 1.38 head remaining at the end of pipe 1 will therefore be the head available for pipe 2 and this figure is entered in column 6 for pipe 2. This head remaining at the end of pipe 1 will also be available for pipe 7.

Similarly, for pipe runs 2, 3 and 4, taking the permissible head loss value of 0.09 from Fig. 51, the required pipe sizes are 28, 28 and 18 mm respectively and the actual rate of head losses 0.06, 0.04 and 0.06. From these figures the head used and head remaining are calculated and tabulated. To calculate pipe sizes for pipe runs 5 and 6, tabulate pipe numbers, design flow and actual length. Assume pipe diameters of, say, 18 and 15 mm respectively, as these are common pipe sizes for branches to baths and WCs; then calculate equivalent lengths for fittings as before, based on these assumed pipe sizes, and tabulate total equivalent lengths. The head available for pipe 5 is taken from Fig. 51 as the head remaining at the end of the pipe run 3, that is 1.14. From the head available and the total equivalent length a permissible head loss value of 0.16 is calculated and a pipe size of 22 mm is taken from Fig. 51. The head available for pipe run 6 is the head remaining at the end of pipe run 2, that is 1.22, less half a metre, the height of the top of pipe run 6 above that of pipe run 4. The head available is therefore 0.72, the permissible head loss value is 0.12 and the pipe size is 15 mm.

To determine the pipe size for pipe run 7 tabulate in columns 1, 2 and 3 as before. Now assume a pipe size for pipe run 7. As the flow in 7 is less than in pipe 1 and the head will be greater, assume that pipe run 7 will be the next size smaller than 1, that is 35 mm. On this assumption calculate equivalent length for fittings of 1.4 and total equivalent length of 4.4. The head available at the function of pipe runs 1 and 7 is the head at that point: 2.5 less the head used in pipe 1, which from Fig. 51, column 10, is 0.62. Therefore the head available at the junction of pipes 1 and 7 is 2.5 − 0.62, which is 1.88. The head along the length of pipe 7 is 3.0 so the head available in pipe run 7 is 1.88 + 3.0, or 4.88. The permissible head loss value is the head available divided by the total equivalent length, that is 4.88 over 4.4 = 1.1. From Fig. 50, for a rate of flow of 1.42 and a permissible head loss value of 1.1, a 28 mm pipe is selected. The actual head loss is then as before and the head used and head remaining are calculated and tabulated. The head remaining will then be tabulated as available for pipe run 8.

The procedure outlined above is used in selecting pipe sizes for the rest of the pipe installation.

The table is a record which can be used to check the calculations leading to the selection of pipe sizes, to confirm that actual head losses do not exceed the permissible head losses, and therefore that pressure is available to provide the required rates of flow at taps and as a basis for subsequent calculations required by any change of plans.

The calculations shown illustrate the method used to determine pipe sizes required to provide a reasonable rate of flow of water from taps. To reduce the labour of manual calculation there are various computer programs that will undertake the necessary calculations once the basic information and assumptions have been supplied.

In the example of selection of pipe sizes for the installation shown in Fig. 49, a possibility was assumed that all the taps might be open at the same time and pipe sizes were selected for this. For small pipe installations, such as those for houses and other small buildings, and

Table 4 Loading units.

Appliances	Loading units
WC flushing cistern (9L)	2
Wash basin	$1\frac{1}{2}$ to 3
Bath tap of nom. size $\frac{3}{4}$	10
Bath tap of nom. size 1	22
Shower	3
Sink tap of nom. size $\frac{1}{2}$	3
Sink tap of nom. size $\frac{3}{4}$	5

for branches from main pipe runs in large installations, it is usual to assume pipe sizes sufficient for simultaneous use of all taps. In these situations only small pipe diameters will be required and there would be no appreciable economic advantage to reduction of pipe sizes by making another assumption.

In extensive pipe installations it is usual to assume a frequency of use for the taps to sanitary appliances so that smaller pipe sizes may be used than would be were it assumed that all taps were open simultaneously. Frequency-of-use values for individual sanitary appliances are expressed as loading or demand units, as set out in Table 4. The total of these units for sanitary appliances is used to determine notional rates of flow in pipes. The loading units of all the sanitary appliances shown in Fig. 49 are:

- 3 WCs, $3 \times 2 = 6$
- 7 basins, $7 \times 11/2 = 10/1/2$
- 2 baths, $2 \times 10 = 20$
- total of $36\frac{1}{2}$ loading units

This total of loading units would require a rate of flow of 0.68 litres per second in pipe run 1. Applying this figure to Fig. 50, with a permissible head loss figure of 0.09 in pipe run 1, a pipe size of 28 mm would be selected – compared with the 42 mm pipe based on an assumption of simultaneous use of all taps.

If the same loading units are then applied to the pipe runs 2, 3 and 4, the sum total of the units is so small as to make no significant difference in the selection of pipe sizes and the sizes previously selected will be used. From this example it will be seen that the use of loading units to determine rates of flow in pipes makes no significant economy in the selection of pipe sizes in pipe runs that serve a few sanitary appliances. As a general rule where pipe runs serve fewer than say ten sanitary appliances it is not worth using loading units to economise on pipe sizes.

2: Sanitary Appliances

Sanitary appliances, sometimes termed sanitary fittings, include all fixed appliances in which water is used either for flushing foul matter away or in which water is used for cleaning, culinary and drinking purposes. The former, termed soil appliances, include WCs and urinals, the discharge from which is described as soil, or soiled or foul water. The second type, termed waste appliances, includes wash-basins, baths, showers, sinks and bidets, the discharge from which is described as waste water.

The reason for the distinction between WCs and urinals as soiled water appliances and the others as waste water appliances, comes from the period before the construction of sewers in the mid nineteenth century. Before then, soiled or foul water appliances drained separately to cesspits or cesspools and waste water to soakaways. Cesspits and cesspools were at intervals cleared of the solid matter that had settled to the bottom and this decomposing matter was spread over land as a form of manure.

Today the soiled and waste water discharge from all sanitary appliances is discharged to a common sewer in most urban areas, and to cesspools or sewage treatment plant in outlying areas. There is no good reason today to differentiate between soiled and waste water fittings on the grounds of separation of discharges.

SOIL APPLIANCES

WC suite

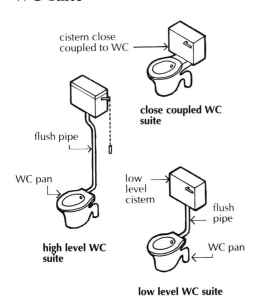

Fig. 52 WC Suites.

The majority of WCs today are sold as a matched set of WC pan, seat, flushing appliance and any necessary flush pipe, which together are described as a WC suite.

The letters WC stand for water closet, the word closet referring to the small room, enclosure or closet in which the early soiled water pans were enclosed when they replaced the original earth closets.

A WC pan is a ceramic or metal bowl to take solid and liquid excrement, with an inlet for flushing and a trapped outlet. The seat is usually a plastic ring secured to the back of the pan. The usual flushing appliance is a cistern designed to discharge water rapidly into the pan through a flush pipe, for cleaning and disposal of contents.

The flushing cistern may be fixed high above, near to or closely coupled to the pan, the three arrangements being described as high-level, low-level (low-down) or close-coupled WC suites, as illustrated in Fig. 52. The high-level suite is no longer popular because of the long unsightly flush pipe and the noisy operation of the flush. It has the advantage that the force of water from the long flush pipe effectively cleans the pan. The low-level or low-down suite is much used today for its appearance. The flush is less noisy than that of the high-level suite, but less effective in cleansing the pan.

The close-coupled suite is so called because the flushing cistern is fixed directly above the back of the pan for the sake of appearance. As the flush water does not discharge into the pan with force, it is at once comparatively quiet in operation and less effective in cleansing. A siphonic WC pan is generally part of a close-coupled suite for the sake of its comparatively quiet operation. Close-coupled suites are more expensive than low-level suites.

WC pans

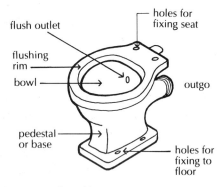

Fig. 53 Pedestal WC pan.

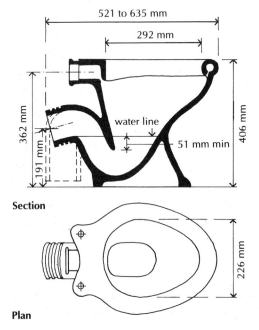

Fig. 54 Wash down WC pan.

Washdown WC pan

Most WC pans are of the pedestal type, the base or pedestal being made integral with the pan. The pan is secured to the floor with screws through holes in the pedestal base to timber plugs in solid floors or directly to timber floors. Fig. 53 shows a typical pedestal WC pan.

The flushing rim is designed to spread the water, which discharges through the flush outlet, around the pan to wash down the sides of the bowl. In hard water areas the rim may become coated with lime scale and need fairly frequent cleansing. Spigot end connections for the flush pipe and for the waste pipe are moulded integrally with the pan.

Most WC pans are made of vitreous china which, after firing, has an impermeable body and a hard, smooth, glazed finish which is readily cleaned. The glazed finish to pans is generally white but may be finished in various pastel colour glazes for appearance sake.

The flushing cistern body and cover to close-coupled WC suites is also made of vitreous china for the sake of appearance. WC pans have integral traps to contain a water seal against odours from the drain pipes or drains.

The two types of pan in use today, the washdown and the siphonic, are distinguished by the operation of the flush water in cleansing and discharging the contents of the pan. In the washdown pan the flush water runs around the rim to wash down the bowl and then overturns the water seal to discharge the contents. In the siphonic pan the flush water washes the sides of the bowl and also causes a water trap or traps to overturn and create a siphonic action which discharges the contents. The purpose of this arrangement is to effect a comparatively quiet flush and discharge of contents.

The flush of water discharged from the old high level and the current low level flush cisterns is generally more effective in cleansing the sides of the WC bowl than is the discharge of flush water from a close-coupled WC suite.

The flush water discharges into the washdown pan around the rim to cleanse the sides of the pan, and as the bowl fills the water in the trap overturns to discharge the contents and as the flush continues it refills the trap.

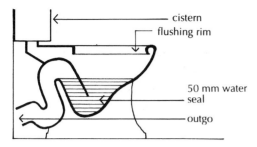

Fig. 55 Single seal siphon WC pan.

The trap is the 51 mm minimum projection of the pan into the water lying in the base of the pan. This acts as a seal against any foul smells that might otherwise rise from the drains. A minimum depth of water seal is set to allow for evaporation of water.

Fig. 54 is an illustration of a typical washdown pan. The back of the pan is near vertical and the sides slope steeply to minimise fouling. In the design of the pan the area of water in the pan should be as large as practical to receive foul matter.

The discharge of flush water down the flush pipe from the traditional high level WC suite is particularly noisy and as the flush water descends with some force there is a likelihood of some water spillage outside the pan. The discharge of flush water from the low level WC suite is both less noisy and less liable to spillage.

Siphonic WC pan

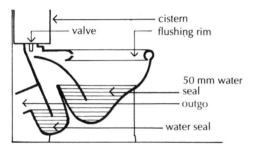

Fig. 56 Double seal siphonic WC pan.

These WC pans effect a discharge of the water in the bowl by the dual action of the flush water and a siphonic action caused by a momentary reduction in pressure that siphons out the liquid contents.

In the single seal pan illustrated in Fig. 55 the flush water causes a full bore flow in the outgo which creates a momentary reduction in pressure which assists the discharge by siphonic action. In the double seal pan illustrated in Fig. 56, some of the flush water enters the air space between the two water seals and this causes the lower water seal to overturn; the momentary reduction in pressure causes the upper water seal to be siphoned out and so discharge the contents of the WC bowl. In both the single seal and the double seal pans the continuing discharge of flush water refills the water seals that have been previously discharged.

It is because the flushing cistern is fixed close to the pan in close-coupled suites, that the discharge of flush water is not particularly vigorous, and the siphonic action pans are used to augment the normal flushing action. Because of the more complex construction of the discharge outgo of the siphonic pans they are more liable to fouling in use and more difficult to clean than a more straightforward washdown pan.

Discharge outgo

The majority of washdown WC pans are made with an outgo that is near horizontal, with a small slope down as illustrated in Fig. 54. This standard arrangement is used for simplicity in production and consequent economy. This type of outgo, described as a P trap outgo (Fig. 57), suits most situations as drain fittings are available to provide a connection to soil pipes relative to the position of the WC.

In some situations the WC pan may discharge to a drain below the floor level and it is convenient to have a vertical outgo. This type of

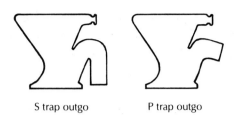

S trap outgo   P trap outgo

Fig. 57 WC pan outgoes.

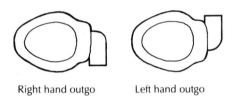

Right hand outgo   Left hand outgo

Fig. 58 WC pan outgoes.

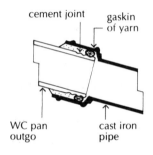

cement joint   gaskin
                  of yarn

WC pan   cast iron
outgo   pipe

Fig. 59 WC to cast iron pipe.

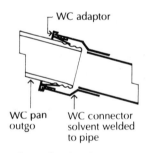

WC adaptor

WC pan   WC connector
outgo   solvent welded
                  to pipe

Fig. 60 WC to plastic pipe.

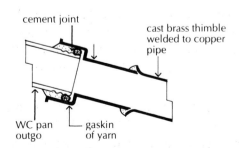

cement joint   cast brass thimble
                          welded to copper
                          pipe

WC pan   gaskin
outgo   of yarn

Fig. 61 WC to copper pipe.

outgo is described as an S trap (Fig. 57). P traps and S traps are so named for their sectional similarity to the letters.

Where a WC pan has to discharge to one side of its position, rather than straight back through a wall, it may be convenient to have a left hand or right hand outgo rather than several unsightly drain fittings. This somewhat exceptional arrangement is illustrated in Fig. 58. The hand, either right or left, is indicated by facing the front of the pan.

The connection of the outgo of a ceramic pan to the branch drain pipe is a common site of leaks because of the difficulty of maintaining a watertight joint and making some allowance for inevitable movement between the pan and the drain branch. In use the pan is designed to support the weight of the user with consequent inevitable movements of the pan which are more pronounced where the pan is fixed to a timber boarded floor.

The traditional joint between the spigot, open pipe, end of the WC outgo and a cast iron drain branch is a cement joint as illustrated in Fig. 59.

The junction of the spigot end of the outgo and the collar of the cast iron pipe is filled with a gaskin of yarn rammed firmly into place to prevent cement running into the pipe, and to align the spigot with the collar. Wet cement and sand is then rammed into the joint. The advantage of this joint is that it is comparatively simple to make. Its disadvantages are that as the cement dries it will shrink and may, if too wet a mix, crack, and this rigid joint may cause the spigot end of the outgo to crack due to movement.

An alternative to the cement joint has been a red lead putty joint. A mixture of linseed oil putty and red lead paint is rammed into the joint on to a gaskin of yarn. The advantage of this joint is that the red lead putty is sufficiently plastic to take up small movements. The disadvantage of the joint is that the material takes some time to harden.

The majority of soil pipes and branches now used are run in plastic pipe sections. The connection of the ceramic pan outgo to the plastic branch is effected by a plastic connector which is solvent welded to the soil pipe branch and whose socket end fits around the pan spigot. The seal is made with a tight fitting plastic adaptor that fits around the pan spigot outgo and adaptor as shown in Fig. 60.

The connection of a ceramic WC pan outgo to a copper branch pipe is made with a cast brass thimble. The thimble has a socket end to fit around the spigot end of the WC pan outgo. The plain end of the thimble is welded to the belled out end of the copper branch pipe. The joint between the socket end of the thimble and the spigot end of the WC pan outgo can be made with a cement and sand joint, a red lead putty joint or one of the proprietary jointing compounds packed into the joint against a gaskin of yarn. This joint is illustrated in Fig. 61.

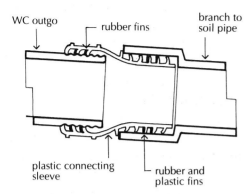

Fig. 62 Plastic sleeve connector for WC pan outgo.

Plastic connecting sleeves specifically made for the connection of the spigot end of a WC pan outgo and the socket ends of branch drain pipes are manufactured from plastic. The ends of the connector are formed with annular rubber and plastic fins to act as water seals as illustrated in Fig. 62.

The connecting sleeve is first forced into the socket end of the branch drain pipe so that the protruding rubber and plastic fins make a tight fit. The spigot end of the WC pan outgo is then worked into the other end of the connecting sleeve with such careful easing as is necessary for the protruding fins to make close contact with the outgo. Because these sleeve connectors are a necessarily tight fit, careful effort is required to fit them into place to make a watertight joint without deforming the connector. The advantage of these connectors is that they provide sufficient flexibility to accommodate any slight movement between the pan and the drain.

WC seats

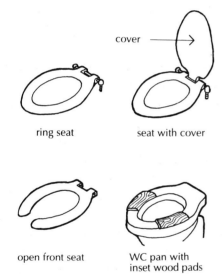

Fig. 63 WC seats.

The usual WC seat is in the form of a moulded plastic ring that fits the top of the WC pan. The back of the seat is bolted to pillars and a rod so that the seat is secured in position and can be lifted. Commonly a separate lift-up cover is fixed to the rod and pillars so that it can be raised or lowered to serve as a seat.

The ring seats and lift-up covers in Fig. 63 are moulded from plastic and finished in a small range of colours, some of which are chosen to match coloured ceramic WC suites.

The open fronted lift up seat illustrated in Fig. 63 has been used in male toilets to minimise fouling. As the effectiveness of this type of seat depends on sensible use – a rare commodity – it is less used than it was.

The inset wood pads fixed to one WC pan in Fig. 63 have been used in communal toilets to avoid the damage that occurs to lift-up seats in general use. Because of the careless use and the difficulty of thoroughly cleaning the wood pads, this arrangement is no longer much used.

Flushing cisterns

With the increasing consumption of domestic water and limited supplies, the water authorities have of recent years amended regulations set out in Water Byelaws to limit water use. In the average household up to 40% of total water consumption is by the use of WC flushing cisterns, which until recently discharged 9 litres of water with each flush.

Before the current byelaws were implemented, a dual-flush cistern was introduced. It was designed to discharge a 4.5 litres half flush by a partial press of the operating lever and a full flush through a full press of the flushing lever. It required some care in use to be effective due to the difficulty of distinguishing between a half and a full press on the flush lever, with the consequence that a full flush was normal.

The current Water Supply Byelaws, that came into effect in 1988, gradually phased out cisterns designed to give a 9 litre flush in favour of cisterns giving a 7.5 litre flush, effective from January 1991 in England and Wales and January 1993 in Scotland for all new installations. Replacement of cisterns of up to 9.5 litres will be permitted. Dual flush cisterns are no longer accepted.

Some water authorities in Europe accept the use of flushing valves that operate to discharge a predetermined volume of water by the operation of a lever or a push. With wear these valves may discharge more water than they were originally designed to discharge and with wear they may drip and waste water. Because of this the water authorities in this country insist on the use of flushing cisterns, which used to be called WWPs – water waste preventers.

Flushing cisterns are made of enamelled or galvanised pressed steel, of plastics or vitreous china. High-level and low-level cisterns are usually of pressed steel or plastic and close-coupled of vitreous china to match the material of the WC pan. Galvanised steel cisterns are used for fixing inside ducts behind the WC pan.

The WC flushing cistern illustrated in Fig. 64 is used for surface fixing for low level WC suites. The cistern body is either made from glazed ceramic, finished in white or a limited range of pastel colours to match coloured WC pans, or in plastic for economy. The cistern body is made in two pieces: the body and a lift-up cover or lid for access to the flush apparatus inside.

Perforations for water supply, overflow pipe, operating lever or push, and for a flush pipe, are provided. These cisterns, particularly the ceramic type, are somewhat bulky. The cistern is secured to a wall by two screws or bolts through the back of the body, above the water line, into plugs in the wall.

The so-called 'slimline' cistern, illustrated in Fig. 65, is designed for use with low level WC suites. It is wider and taller than the cistern in Fig. 64, so that it does not protrude from the wall too much. This is an advantage from the point of view of appearance and because the pan does not have to project so much into the room. A disadvantage of the bulky cistern in Fig. 64 is that the pan has to be fixed some distance forward, if the lift-up cover and seat are to remain in place when lifted.

The 7.5 litre slimline cisterns are usually made from rigid polystyrene plastic with a separate body and cover. The body is holed for the lever, water supply, overflow and flush pipe. The slimline cistern in Fig. 66 is designed for fixing in a duct or space behind WC pans, where it is concealed from sight with the operating lever protruding through a thin cover panel. As the cistern is out of

7.5 litre rigid plastic WC cistern for surface fixing

Fig. 64 WC cistern.

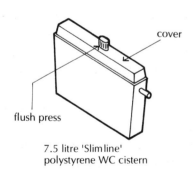

7.5 litre 'Slimline' polystyrene WC cistern

Fig. 65 Slimline cistern.

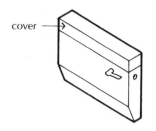

Fig. 66 7.5 litre 'Slimline' cistern.

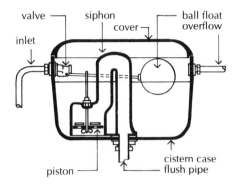

Fig. 67 7.5 litre WC flushing cistern.

Small bore macerator sanitary system

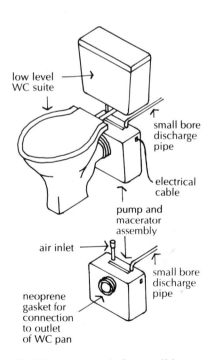

Fig. 68 Macerator unit for small bore discharge for WCs.

sight it is usually made of galvanised steel or rigid plastic for economy. The water service and overflow pipes are out of sight and the flush pipe runs from behind the cover panel out to connect to the back of the pan.

Flushing cisterns discharge water in one operation through a flush pipe or directly to the pan by siphonic action. Fig. 67 is an illustration of a typical flushing cistern. It will be seen that the cistern is filled through a valve operated by a ball float and arm similar to that described for water storage cisterns. The Water Supply Byelaws require that there be an air gap between the highest level of water in the cistern and the outlet of the float valve. This air gap is related to the bore of the supply pipe and is the same as that for storage cisterns.

The plastic siphon is operated by a lever which raises a piston to force water over the siphon bend and the siphonic action causes the water in the cistern to follow, through perforations in the piston, up the siphon and down the flush pipes in one go. A distributing or cold water supply pipe is connected to the valve of the cistern and an overflow warning pipe is run outside the building or to discharge over the pan or onto the floor where the WC is fitted.

The Building Regulations permit the use of small bore pipe discharges from WCs. They have been used in Europe for many years and their use depends on the macerator and pump fitted to the outlet of WC pans The electrically powered macerator and pump come into operation as the normal flush of a WC pan, by a conventional cistern, fills the pan. A macerator is a rotary shredder whose blades rotate at 300 rpm and reduce solid matter to pulp, which is, with the flush water, then pumped along a small (18 to 22 mm) pipe to the discharge stack. The macerator (shredder) and pump unit, which is about $340 \times 270 \times 165$ mm, fits conveniently behind a WC pan (Fig. 68). The unit is connected to the horizontal outlet of a BS 5503 pan and a small bore outlet pipe. The macerator and pump are connected to a fixed, fused electrical outlet.

The particular advantage of the small bore system is in fitting a WC in either an existing or a new building some distance from the nearest foul water drainage stack, with a small bore (18 to 22 mm) pipe that can be run in floors or can be easily boxed in. In addition, because of the pumped discharge, the small bore branch discharge pipe can carry the discharge for up to 20 m with a minimum fall of 1 in 180 and can also pump the discharge vertically up to 4 m, with a reduced horizontal limit, which is of considerable advantage in fitting WCs in basements below drain levels. The macerator and pump unit can also be used to boost the discharge from other fittings such as baths, basins, sinks, bidets and urinals along small bore runs, with a minimum fall, and over considerable runs not suited to normal gravity discharge.

The small bore discharge system is not a substitute for the normal short run branch discharge pipe system for fittings grouped closely around a vertical foul water drainage system, because of the additional cost of the macerator and pump unit and need for frequent periodic maintenance of the unit.

Urinals

The three types of urinal in general use are the stall urinal, slab urinal and bowl urinal.

Stall urinals

The stall urinal consists of heavy, individual stoneware stalls with either a salt glazed or white glazed finish, each stall having its own integral channel. The stalls are set in place on a solid floor, against a wall. The junction between individuals stalls is covered with salt glazed or white glazed rolls or facing pieces, to serve as a finish to the joint between the stalls and also to afford some privacy to users.

The channel, at the foot of the stall, drains to a brass outlet to a branch drain pipe. The floor finish overlaps the edge of the channel. The three stall unit illustrated in Fig. 69 comprises two end stalls and one plain stall. An automatic flushing cistern, mounted on the wall, flushes through a horizontal pipe to spreaders to each stall. The stalls and the facing pieces are bedded in cement and sand, and joints are finished in cement.

This heavy, robust type of urinal was much used in Victorian times for the many public lavatories that were a feature of public utilities in towns. The sturdy materials were used for their resistance to vandalism, but this type of urinal, which takes up space and is laborious to clean, is less used than it was.

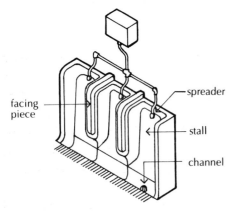

Fig. 69 Stall urinal.

Slab urinals

Slab urinals, which are of less heavy construction, have largely taken over from stall urinals for use in public lavatories. These urinals consist of flat, white glazed ceramic slabs and white glazed ceramic channels that are bedded in cement and sand against a solid wall, with projecting end slabs to each range of urinals as illustrated in Fig. 70. The joints between the slabs and slabs and channels are pointed in cement. An automatic flushing cistern discharges water to a sparge pipe fixed over the slabs. The slabs are flushed by water from perforations in the sparge pipe. This straightforward, economical type of urinal is used for public lavatories and in schools as it is reasonably vandal proof, except for the sparge pipe which seems to be a favourite trophy for many.

The plain slab urinal, which is comparatively cramped when in full use, is avoided by some owing to lack of privacy because they prefer to

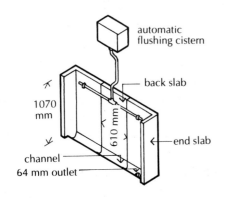

Fig. 70 Slab urinal.

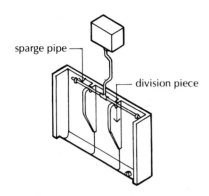

Fig. 71 Slab urinal with divisions.

Bowl urinals

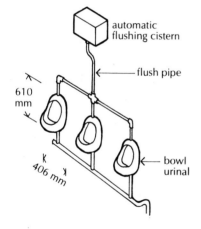

Fig. 72 Bowl urinal.

use an enclosed WC closet if possible. To provide a degree of privacy, slab urinals with individual division pieces between the stalls are often installed; glazed division pieces are bedded in position between the slabs, as shown in Fig. 71. The division pieces, which may be damaged and become loose in time, to some extent add to the sense of being cramped when the urinal is in full use. An automatic flushing cistern flushes the stalls through a sparge pipe fixed to the range of stalls.

Slab and stall urinals are best finished with the floor finish run up to and over the edge of the channel to facilitate washing floors into the channel. On upper floors these urinals are formed with a step up to them to avoid the drain and gully protruding down through the ceiling below. Such step-ups can be a hazard to the unwary who step back without looking and may stumble or fall.

Individual white glazed ceramic bowl urinals fixed to a wall are used for ease of cleaning and the sense of space they give. For privacy these bowl urinals are separated by division pieces fixed to the wall between them. The disadvantage of these bowl urinals is that the floor over which they are fixed fairly readily becomes fouled by careless use of the bowl and needs frequent washing, and the bowls and flush pipes may be fairly readily damaged by careless use or vandalism.

The bowls, which should be bolted to a wall or support, are bedded in cement and sand and the joints finished with one of the silicone sealing compounds designed for the purpose. The automatic flushing cistern, flush pipe and waste, may be fixed into the wall as illustrated in Fig. 72, or fixed behind a partition or panel framing to which the bowl urinals are fixed.

Urinals are flushed by automatic flushing cisterns fixed above the urinal and discharging through a flush pipe, spreaders or sparge pipe. The automatic flushing cistern is of 4.5 litres capacity per slab, stall or bowl and the cisterns are adjusted to flush every 20 minutes.

The Water Supply Byelaws limit the flush of cisterns to two or more urinal bowls or stalls or two or more widths of slab each exceeding 700 mm in width, to 7.5 litres per hour for each unit and 10 litres per hour for a single urinal bowl or stall.

The cistern is filled directly from the distributing pipe, the rate of filling and therefore the frequency of flush being controlled by a valve. When full the siphon overturns and discharges the contents in one go. The flush from the cistern down the flush pipe is then distributed over the urinal by the individual spreaders to each slab, stall or bowl or by means of a perforated pipe, termed a sparge pipe to slab urinals only.

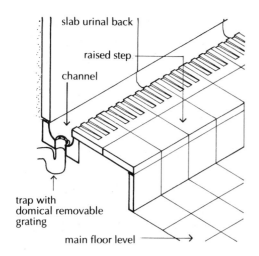

slab urinal back

raised step

channel

trap with
domical removable
grating

main floor level

Fig. 73 Urinal step.

One outlet to trap and branch discharge pipe is used for up to six stalls or slab units, the outlet being in the channel to the slab or in the channel of one of the stall units. The outlet, 40 mm minimum diameter, is covered with a domed, gun metal grating and the outlet connected to a trap and waste. Deposits build up rapidly in the trap of urinals and traps of glazed ceramic, vitreous enamel or lead are often used to take the corrosive cleaning agents used.

To accommodate the channel of urinals in a floor, a step is often formed. Fig. 73 shows a typical urinal slab with channel, trap, and waste.

Bowl urinal outlets are often connected to a combined waste with a running trap.

WASTE WATER APPLIANCES

Waste or waste water appliances include basins, baths, sinks and bidets.

Wash basins

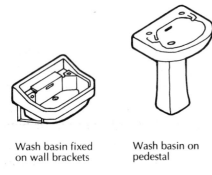

Wash basin fixed
on wall brackets

Wash basin on
pedestal

Fig. 74 Wash basins.

Wash basins, designed for washing the upper part of the body, are supported by wall brackets or by a pedestal secured to the floor, as shown in Fig. 74. The standard wash basin consists of a bowl, soap tray, outlet, water overflow connected to the outlet, and holes for fixing taps.

The usual wall-mounted basin is fixed on enamelled cast-iron brackets screwed to wood plugs in the wall. The more expensive pedestal basin consists of a basin and a separate vitreous china pedestal that is screwed to the floor and on which the basin is mounted. The purpose of the pedestal is to hide the trap, waste and hot and cold service pipes. Either the whole or a large part of the weight of the basin is supported by the pedestal which should preferably be fixed to a solid floor to provide solid support. A resilient pad is fitted between the bottom of the basin and the top of the pedestal as the two separately-made fittings rarely make a close fit.

The majority of wash basins are made of vitreous china. A wide range of sizes and designs is available, ranging from small corner basins and hand basins to basins large enough for bathing a small child. In recent years plastic basins have been made which are suited in particular to fitting to stands and working tops.

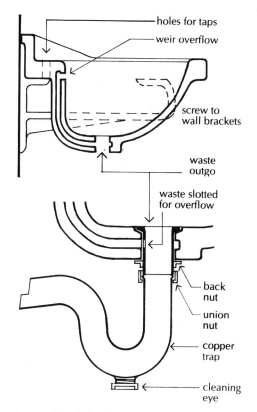

Fig. 75 Wash basin waste and trap.

Baths

Fig. 76 Magna square.

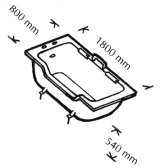

Fig. 77 Baths.

Hot and cold pillar taps connected to 12 or 15 mm hot and cold distributing or supply pipes are fixed to or over wash basins. To prevent the possibility of foul water in the basin being back siphoned into the pipes, it is a requirement of the Water Supply Byelaws that there be an air gap between the outlet of the taps and the spill over or top edge of the basin, as described in Chapter 1.

To avoid an overflow of water from a basin, a weir overflow is usually formed during the manufacture of basins. This consists of a hole in the top of the basin, which can drain to the outlet, as shown in Fig. 75. These weir overflows will become fouled with soap and hairs in time and become blocked. As a blockage is difficult to clear it is wise to clean the overflow from time to time with soda and hot water.

A waste outgo, with slot to drain the weir overflow, is formed in the basin. A waste is fitted to the outgo, bedded in setting compound and secured with a back nut, as shown in Fig. 75. A copper or plastic P trap with 75 mm water seal is then connected to the waste outgo and the copper or plastic waste pipe.

When the basin is drained of water, the waste pipe may run full bore at the end of the discharge, cause self-siphonage and empty the water seal in the trap. Because of the shape of the basin there may not be sufficient tail-off water to refill the trap. To avoid this the waste pipe should be of limited length or an air-admittance valve should be fitted.

The type of bath most used is the standard Magna square ended bath, illustrated in Fig. 76. These baths are made of porcelain-enamelled cast iron, or enamelled pressed sheet steel or plastic. The enamel finish of the heavier and more rigid cast iron bath is less likely to be damaged than that on the lighter pressed steel bath. The plastic bath does not have the hard bright finish of the metal bath and is light-weight and not liable to rust, but it is fairly readily scratched. Baths are finished in white and a limited range of pastel colours.

The modified Magna bath has a drop edge side for ease of entry to the bath and a chromium plated handle for convenience of getting out of the bath, as shown in Fig. 77.

The Magna bath has a rectangular profile rim designed to accommodate end and side panels, and an outlet, overflow, holes for taps and adjustable feet, as shown in Fig. 77. The square ends are designed for fixing against a wall or partition, with one side and two end panels fitted to timber bearers under the rim of the bath, or for fitting into a purpose-made recess with one side panel only.

The traditional roll top or tub bath (Fig. 78), made of enamelled cast iron, was for years the usual form of bath. It was free sanding or set against a wall or partition. As the rounded end did not provide a ready means of fixing side and end panels, this type of bath was

Fig. 78 Tub bath.

Fig. 79 Sitz bath.

sometimes totally enclosed in timber framed panelling with a removable top, in kitchens. Because of the cast iron body and heavy enamel finish these baths are very robust and durable.

This type of bath has regained favour and is often fitted into new and refurbished homes as a feature, often free standing as a centre piece to a large bathroom.

Sitz or sitting baths have a stepped bottom to form a seat, as illustrated in Fig. 79. These cast iron enamelled baths were originally made for the elderly or infirm who had difficulty in climbing into and out of a normal bath and could manage bathing in a sitting position. These baths, which have been used where space is restricted, have by and large been replaced by showers.

Baths are fixed on adjustable feet against a wall or in a recess with side and end panels of preformed board, plastic or metal secured to wood or metal brackets or frames. The plaster or tile wall finish is brought down to the top of the bath rim. In time, particularly on timber floors, the joint between the top of the bath and the plaster or tile will crack.

The discharge from a bath is unlikely to run full bore and cause self-siphonage and loss of water seal to the trap, and if the trap were to lose its seal by siphonage it would be filled by the tail-off water from the flat bottom of the bath. The length and slope of a bath waste is therefore not critical. A 75 mm seal trap is fitted to the bath waste and connected to the 40 mm waste.

An overflow pipe is connected to the bath and either run through an outside wall as an overflow warning pipe or connected to the bath outlet or trap, as illustrated in Fig. 80. Overflow warning pipes through an outside wall should have hinged flaps at the end, otherwise an appreciable cold draught may blow into the bath.

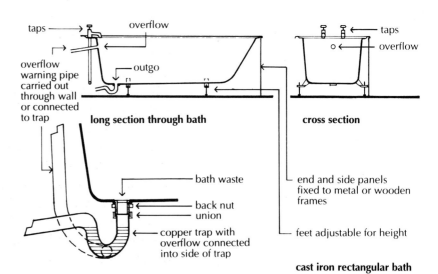

Fig. 80 Plumbing to bath.

The 18 mm cold and hot distributing or supply pipes are connected to either individual pillar taps, a mixer with taps or a shower fitting, or both, with an air gap between the outlet and spillover level of the bath, similar to that for basins.

Where shower fittings are provided, the wall or walls over baths should be finished in some impermeable material such as tile, and a waterproof curtain be provided.

Showers

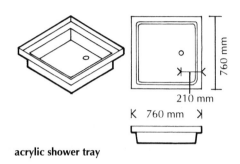

acrylic shower tray

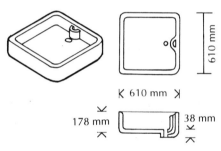

glazed fireclay shower tray

Fig. 81 Shower trays.

Sinks

Fig. 82 Belfast sink.

The conventional shower or shower bath consists of a shower tray or receiver of glazed ceramic, enamelled cast iron or plastic to collect and discharge water, with a fixed or hand-held shower head or rose and mixing valves. The shower is either fixed in a wall recess or may be free standing with enamelled metal or plastic sides. The walls around fixed showers are lined with some impermeable material such as tile, and the open side is fitted with a waterproof curtain. Fig. 81 shows some shower trays. The tray with a waste and no overflow is for use as a shower only. The tray with a waste plug and an overflow is for use either as a shower or a foot bath.

A shower compartment is often surrounded with an upstand curb or may be sunk into the floor to contain the shower water that would otherwise spill over the surrounding floor. The shower tray may be fixed onto or recessed into the floor. A 75 mm seal trap is fitted to the tray and connected to a 40 mm waste. The discharge from a shower tray is unlikely to run full bore down the waste and as there is little likelihood of self-siphonage the length and slope of the waste are not critical.

The Water Supply Byelaws require a double check valve assembly to both hot and cold supply pipes to showers, where the head can be lowered below the spill over level of appliances.

The traditional sink was made of glazed stoneware, usually white glazed inside the bowl and salt glazed outside. They were heavy, durable sinks with a capacious bowl requiring substantial support. These highly practical sinks, commonly known as Belfast sinks (Fig. 82), which cannot conveniently be housed in the current kitchen units, lost favour until recently because of their appearance. They are back in fashion partly because the large bowl will accommodate such things as greasy pans that are impossible to clean in the tiny bowls of modern sinks. Fitted under natural teak draining boards these sinks are again in fashion as a feature to modern kitchens.

For many years the Belfast sink was replaced by cast iron enamelled sinks, often with an integrally cast drainer. The solidity of the cast iron base, together with the thick white enamel finish, provided a tough, durable sink that could withstand normal wear and tear

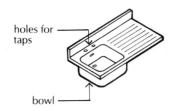

Fig. 83 Single bowl single drainer.

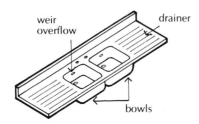

Fig. 84 Double bowl double drainer.

Bidets

Fig. 85 Bidet.

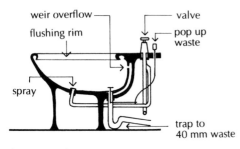

Fig. 86 Bidet.

without damage for many years. The durable white enamel finish did tarnish with use but could be brought back to its original lustrous finish with a little effort. These large, practical sinks have lost favour to the stainless steel sink.

A range of stainless steel sinks, designed to fit into kitchen units, are made with single bowl and drainer (Fig. 83) and double bowl and double drainer (Fig. 84). The sinks are finished in the natural colour of the stainless steel from which they are pressed. The dull grey finish soon loses its initial sheen and is difficult to clean and burnish back to its original finish.

The majority of these sinks have so small a bowl that it is impossible to fully immerse anything larger than a plate for thorough cleaning, presumably on the basis that most people use dishwashing machines.

Most sinks have weir overflows connected to the waste outlet and are holed for fitting hot and cold taps or mixers. A 75 mm seal trap is connected to the sink waste which is connected to a 40 mm copper or plastic waste pipe. Because of the flat base of the bowl a sink waste is unlikely to run full bore and cause self-siphonage.

Bidets, which are appliances for washing the excretory organs, have never been as popular in the UK as they are in the rest of Europe.

A bidet (Fig. 85) consists of a glazed ceramic pedestal bowl which is secured to the floor, usually backing on to a wall or partition. A bidet may be white glazed or finished in a limited range of pastel colours to match other bathroom appliances.

The shallow bowl has a flushing rim, a weir overflow connected to the waste and an inlet for a spray. An optional hand-held spray may be fitted to the hot and cold supply. The bidet operates through the discharge of water around the flushing rim, and a spray of water that rises from the bowl or a hand-held spray. The bowl may be filled with water and drained by the operation of a pop-up waste control. The temperature of the spray water is controlled by hot and cold water valves. Fig. 86 shows a bidet with a 75 mm trap from the waste outgo to a 40 mm waste pipe.

As a precaution against the possibility of contamination of the mains supply from a bidet, particularly through the submerged spray, the Water Supply Byelaws require a separate cistern feed to a bidet or other effective device to sprays and hand-held showers, such as double check valve assemblies. These requirements add considerably to the work and cost of fitting a bidet in this country. The discharge from a bidet is unlikely to run full bore in the waste and cause self siphonage.

TRAPS

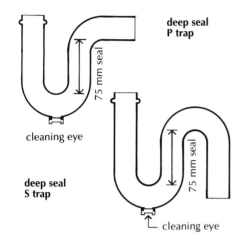

deep seal
P trap

75 mm seal

cleaning eye

deep seal
S trap

75 mm seal

cleaning eye

Fig. 87 Deep seal S and P traps.

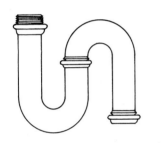

Fig. 88 Two piece copper S trap.

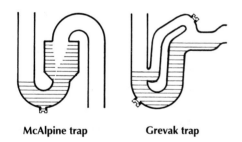

McAlpine trap   Grevak trap

Fig. 89 Resealing traps.

A trap or waste trap is a copper or plastic fitting formed as a bend in pipework to contain a seal of water as a barrier to odours rising from sanitary pipework and drains into rooms. The traps are formed as P or S traps to accommodate the position of the waste pipe relative to the sanitary fitting outlet. At the bottom of the trap is a cleaning eye, which can be unscrewed to clear blockages. Fig. 87 shows P and S traps.

The depth of the water seal is measured from the top of the first bend and the bottom of the second. The traps shown in Fig. 87 are 75 mm deep seal traps – the depth of seal required for all sanitary fittings connected to single stack systems of drainage, except for WCs which have an integral 50 mm seal.

The two piece trap in Fig. 88 is used instead of the one piece trap because it can be adjusted to suit the position of the branch waste pipe relative to the appliance outgo, and can be uncoupled to clear blockages which are common with wash basins and sinks.

Before the single stack system of sanitary pipework was established as the optimum system to control siphonage of water seals, it was common to use resealing traps which contained a reservoir of water. These traps are designed to allow air through the trap or through an air tube to reseal the trap after the main seal has been cleaned by siphonage. The McAlpine trap and the Grevak trap in Fig. 89 work on this principle.

These traps, which are liable to blockages, are little used other than in older installations. Where siphonage may occur, an air admittance valve is used instead of a resealing trap.

3: Sanitary Pipework

PIPEWORK

History

Up to the middle of the nineteenth century few buildings had a supply of water piped above ground floor level. Water closets were either inside or outside at ground level and washing water was carried by jug to basins. From about the middle of the nineteenth century, due to improved pumping techniques, piped water increasingly extended above ground floor level, first to water closets and later to wash basins and baths. This improvement in sanitary and washing facilities was helped by the mass production of sanitary appliances.

The drainage above ground of water closets was usually separate from that of other sanitary appliances. The demands for hygienic arrangements, provoked by the typhus epidemics early in the century, were at first confined to the drainage of water closets. Waste water from basins and baths was not at that time thought to be foul and drains from waste water appliances were often connected to rainwater pipes and soakaways. To avoid drain smell, pipes from sanitary appliances were run directly out of buildings to vertical pipe systems termed stack or stack pipe, fixed to external walls.

A typical arrangement of the drainpipes of this period is shown in Fig. 90. The first-floor WC discharges to a separate, soiled-water pipe connected to the drains. The waste water from the first-floor bath and basin discharges into an open hopper head that also collects the rainwater from the roof, with the hopper head draining to a trapped gully connected to the drains. This was described as the two pipe system.

The advantage of this system is simplicity. The trapped gully taking the discharge of rainwater and waste water acts as an effective seal against odours rising from the drains. As the bath and basin wastes discharge over the hopper head they do not need a trap. By combining rain and waste water the discharge pipe is reasonably flushed.

The one disadvantage of the system is that with careless use the hopper head may become fouled, smelly and eventually blocked. It is a simple matter to clear a hopper head at first-floor level. The system of using hopper heads to collect waste water discharges was also used for buildings over two floors high and it was then that the smells from fouled hopper heads, floods from blocked heads, and the difficulty of clearing hopper heads above ground gave this system a bad name. The use of hopper heads for the collection of waste water has since been abandoned.

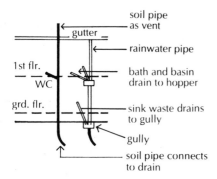

Fig. 90 Two pipe system.

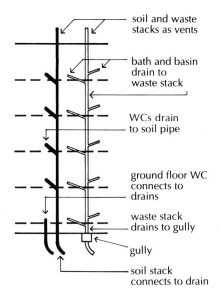

Fig. 91 Two pipe system.

soil and waste
stacks as vents

bath and basin
drain to
waste stack

WCs drain
to soil pipe

ground floor WC
connects to
drains

waste stack
drains to gully

gully

soil stack
connects to drain

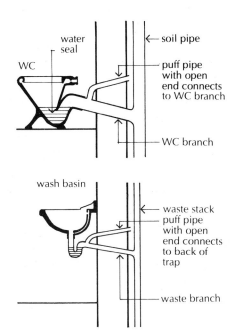

Fig. 92 Puff pipes.

water
seal

WC

soil pipe

puff pipe
with open
end connects
to WC branch

WC branch

wash basin

waste stack
puff pipe
with open
end connects
to back of
trap

waste branch

When the use of open hopper heads lost favour, a two pipe system of separate sealed waste and soil stack systems was adopted, as shown in Fig. 91. There are two separate discharge stacks, one for the WCs and another for waste water discharges. Both pipe systems are sealed and each discharges separately to the drains, the soil stack directly and the waste stack through a trapped gully. In this modification of the original two pipe system, rainwater discharges were separated from the waste water as there were no longer hopper heads to collect the two discharges.

The discharge of water from sanitary appliances causes changes of air pressure in a sealed discharge pipe system. These changes or fluctuations are not in general large or of long duration and it is now known that pressure fluctuations sufficient to unseal water traps can be avoided by careful design of the discharge system. The single stack system in common use today is designed to that end.

When piped water and sanitary appliances were installed above ground, from the middle of the nineteenth century, it was known that there was a likelihood of siphonage of water from traps due to pressure fluctuations. To prevent drain smells, precautions were taken to prevent the siphonage of water from traps by the use of puff pipes and later anti-siphon pipes (vent pipes). The system of anti-siphon pipes became an intricate web of pipes festooned over buildings of the early twentieth century.

To prevent the siphonage or loss of the water seal in a trap it is only necessary to maintain equal air pressure on both sides of the trap to a sanitary appliance. The most straightforward way of doing this is to connect a short length of pipe from the outgo side of the trap to the open air. The short length of pipe was called a puff pipe, presumably because it drew in and expelled a puff of air to stabilise pressure. The use of a puff pipe to the trap of a WC and a basin is illustrated in Fig. 92.

The use of puff pipes to WC pan traps was abandoned for fear of drain smells rising up the soil pipe outside through the puff pipe and into an open window above. It would obviously need a singularly pungent smell propelled by considerable air pressure for this to happen. None the less, the fear of drain smells that has to this day beggared rationalisation of drains, caused the puff pipe to be abandoned. Recently puff pipes have been used to ventilate a discharge pipe system in a building with sealed windows, and the most recent regulations permit the use of air admittance valves, that are a form of puff pipe.

Having condemned the puff pipe, the then accepted means of preventing siphonage of the water seal of traps was the anti-siphon pipe (now called the vent pipe). A separate system of pipes was connected to the trap of all sanitary appliances to equalise air pres-

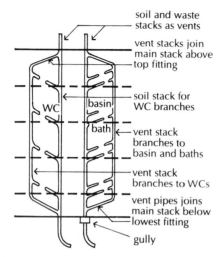

Fig. 93 Two pipe system fully vented.

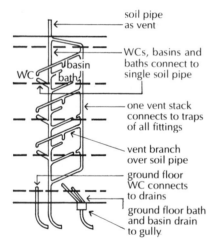

Fig. 94 One pipe system fully vented.

Induced siphonage

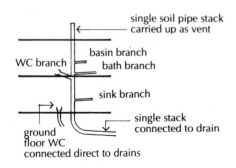

Fig. 95 Single stack system.

sure on both sides of the traps. Fig. 93 shows anti-siphon pipe systems applied to a two pipe system of pipework. There is a stack with branches from the vent pipes connected to the outlet side of each trap to each appliance. This fully vented two pipe system was commonly installed in buildings between the world wars, 1920 to 1939.

This unsightly and uneconomic web of pipes was later modified in the one pipe system which utilised a single discharge stack for the discharge from all appliances. Fig. 94 illustrates a fully vented one pipe system. The one pipe system gave a small economy in pipe runs but did little to improve the unsightly web of pipes on the external face of buildings.

In 1952 the Building Research Station, now the Building Research Establishment, published its first report on sanitary pipework. For some years the station had been carrying out tests in the laboratory and on site to determine the likelihood of siphonage of traps to appliances, with a view to effecting economy in sanitary pipework in housing. The report established that by careful arrangement of the branches to a soil pipe there would be no loss of seal to the traps of appliances and therefore no need for vent pipes.

The recommended single soil pipe with branches from sanitary appliances without vent pipes was called the single stack system (Fig. 95). Since then further work has shown that the single stack system can be used for multi-storey buildings without vent pipes by increasing the size of the stack or by the use of minimal vent pipes with a smaller stack.

From their study of pipe systems in use, the Research Station distinguished three conditions in which there could be loss of water seal to traps of appliances. These three conditions of air pressure fluctuations are:

(1)   Induced siphonage
(2)   Self-siphonage
(3)   Back pressure

Induced siphonage may be caused by a discharge from a WC down a soil stack. As the discharge carries air with it there is a momentary reduction in pressure that may unseal the trap to a branch waste, as illustrated in Fig. 96. WC discharge branches to stacks should be swept in the direction of flow, as illustrated in Fig. 103, to reduce the likelihood of induced siphonage. Induced siphonage may also occur where waste pipes from two appliances connect to a common branch waste pipe.

Self-siphonage

Self-siphonage may occur where the discharge from an appliance runs full bore in the waste pipe at the end of the discharge, causing a reduction in pressure and possible loss of water seal to the trap, as illustrated in Fig. 97. Where loss of water seal to wash basin traps occurs, due to self-siphonage, the trap will not be filled because there is too little tail-off water from the funnel-shaped basin to fill the trap. To reduce the possibility of self-siphonage, wash basin wastes should not be more than 1.7 m long. If the water seal to the traps of baths and sinks is broken by self-siphonage, there is sufficient tail-off water from the flat bottom of these appliances to fill the trap and there is, therefore, less need to limit the length or slope of wastes against self-siphonage.

Back pressure occurs when a discharge reaches the base of a stack at the junction of a branch waste near the base of the stack. The increased pressure caused by the discharge may overturn the water seal in the trap to the branch waste, as illustrated in Fig. 98. To limit back pressure the bend at the base of the stack should have a large radius, shown in Fig. 103.

Similarly, where a branch waste is connected close to a WC branch, a discharge from the WC may cause back pressure in the branch waste. Branch wastes should not be connected to the stack for a depth of 200 mm below the centre line of the WC branch, as shown in Fig. 103.

For economy of sanitary pipework the single stack system, illustrated in Fig. 99, is used today in both domestic and public buildings.

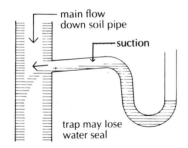

Fig. 96 Induced siphonage.

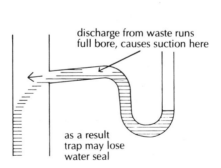

Fig. 97 Self-siphonage.

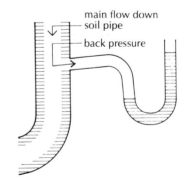

Fig. 98 Back pressure.

Ventilated stack system

Where pressure fluctuations in the stack may be so great as to cause induced siphonage or back pressure, for example in multi-storey buildings, a ventilation pipe connected to the stack is used. This arrangement is the ventilated stack system, shown in Fig. 99.

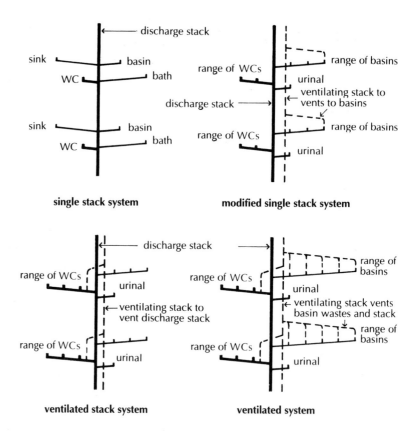

Fig. 99 SIngle stack systems.

Modified single stack system

Ventilated system

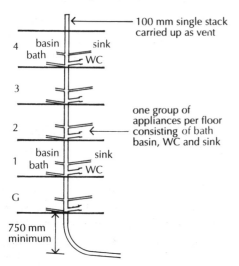

Fig. 100 Single stack for up to five floors.

Where pressure fluctuations in branch wastes to a single stack system may be sufficient to siphon the water seal from traps, a ventilation pipe system is connected to vent the traps. This is the modified single stack system, shown in Fig. 99.

Where pressure fluctuations in the stack and the branch wastes cannot be limited to prevent self-siphonage, induced siphonage and back pressure, for example in multi-storey buildings, a ventilated system is used, as shown in Fig. 99.

The single vertical pipe collecting discharges from all sanitary appliances is the discharge stack and the pipes from all appliances to the stack are discharge pipes. The single vertical ventilating pipe is the ventilating or vent stack and the branches from it to the discharge stack and discharge pipes are ventilating or vent pipes.

Fig. 100 illustrates the application of the single stack system to a five floor residential building with one group of appliances on each floor. The discharge pipes are arranged within the limitations set out in Fig. 103. The 100 mm single stack shown in Fig. 100 can also be used to take the discharge from two groups of appliances per floor for up to five floors.

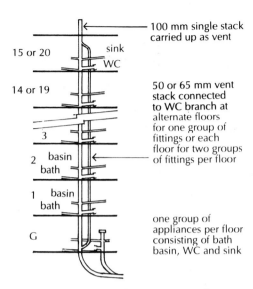

Fig. 101 Single stack for up to 20 floors.

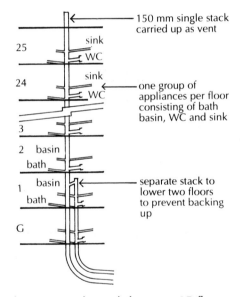

Fig. 102 Single stack for up to 25 floors.

The single stack system of sanitary pipework was originally developed for houses. It has since been developed for use in multistorey buildings, such as flats, where sanitary fittings are closely grouped, floor over floor, so that short branch discharge pipes connect to a common single stack for economy in drain runs.

For blocks of flats of up to five floors, experience has shown that a single stack system of drainage is satisfactory for two groups of appliances on each floor where the fittings are grouped close to the stack. This arrangement may well work satisfactorily for blocks of flats up to say ten storeys providing the length of branches and their entry to the stack conform to the requirements set out.

For blocks of flats 15 to 20 floors high a modified system of the single stack with a ventilating stack can be used. The 50 or 65 mm stack is connected to WC branches on alternate floors for one group of appliances and to WC branches on each floor for two groups of appliances. This arrangement, which serves groups of sink, bath, basin and WC has been successfully used. Fig. 101 is an illustration of this arrangement, where the lowest group of appliances is connected directly to the drains through a substack.

As an alternative, a larger 150 mm single stack may be used without a ventilating stack for one group of appliances per floor. The larger stack limits pressure fluctuation for all but the lower two floors, where pressure is greatest. Fig. 102 is an illustration of this system applied to a 25 storey block of flats. The two lower floors are connected to a separate stack that drains directly to the drains underground.

A condition for the effective operation of the single stack system is that the stack enters the drains underground through a wide sweep bend to limit pressure of discharge as it enters the drain. The modifications of the single stack system have been effectively used where the conditions of close grouping of appliances, short branches and care in arranging the entries to the stack are carefully observed. Departure from these conditions may well result in siphonage or back siphonage.

Building Regulations

Section 1 of the approved document H1 gives practical guidance to meeting the requirements of the Building Regulations for sanitary pipework.

The approved documents give practical guidance to meeting the requirements of the Building Regulations, but there is no obligation to adopt any particular solution given in the documents if the requirements of the Regulations can be satisfactorily met in some other way.

Sanitary pipework includes all pipework used to carry the discharge of 'foul water' from sanitary appliances such as water closets, bidets, baths, wash basins and sinks, and ventilation of pipework as is necessary to prevent foul air from the drainage system entering the building under working conditions.

An acceptable level of performance of sanitary pipework and drainage will be met by any provision of section 1 of approved document H1. To reduce the risk to the health and safety of persons in buildings the foul water drainage system should:

(1)   Convey the flow of foul water to a foul water outfall
(2)   Minimise the risk of blockage or leakage
(3)   Prevent foul air from the drainage system from entering the building under working conditions
(4)   Be ventilated
(5)   Be accessible for clearing blockages

In the document 'foul water' is defined as waste from a sanitary convenience or other soil appliance, and water which has been used for cooking or washing, but does not include waste containing any trade effluent. Foul outfall means a sewer, cesspool, septic tank or settlement tank. The capacity of the pipework, which depends on the size and gradient of the pipes, should be large enough to carry the expected flow, which depends on the type, number and grouping of appliances at any point.

Single stack system

The requirements of the Building Regulations for sanitary pipework can be met by the use of the single stack system of sanitary pipework, which is the system most in use for economy in layout and use of pipework. Fig. 103 is an illustration of a single stack system for a two-storey house, showing a single discharge stack pipe to which are connected branch discharge pipes from a sink, wash basin, WC, bath and ground floor WC.

Traps

Where sanitary appliances discharge foul water to the sanitary pipework system there should be a water seal, provided by means of a trap, to prevent foul air from entering the building under working conditions. There is a water seal trap to each of the appliances. The minimum size and depth of water seal for these traps are set out in Table 5.

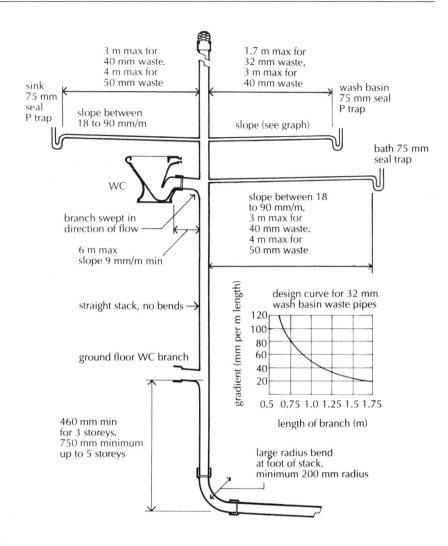

Fig. 103 Single stack system.

Table 5. Minimum trap sizes and seal depths.

Appliance	Diameter of trap [mm]	Depth of seal [mm]
washbasin bidet	32	75
sink bath shower food waste disposal unit urinal bowl	40	75
WC pan (siphonic only)	75	50

Water closet pans have a water seal trap that is integral with the pan in the form of a single or double water seal, as illustrated in Figs. 55 and 56. Baths, bidets, sinks and wash basins have a trap which is fitted to the appliance and connected to the branch discharge pipe. Single and double seal traps are illustrated in Figs 87 and 89. To facilitate clearing blockages there should be a clearing eye or the trap should be removable, as shown in Fig. 87.

To prevent the water seal in traps being broken by the pressures that can develop in a sanitary pipe system, the length and gradient of branch discharge pipes should be limited to those set out in Fig. 103, or a system of ventilating pipes should be used.

Branch pipes

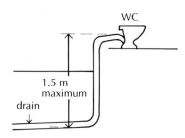

Fig. 104 Direct connection of ground floor WC to drain.

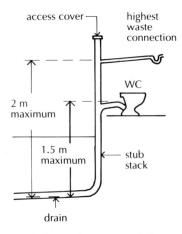

Fig. 105 Stub stack to ground floor appliances.

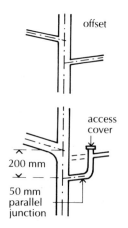

Fig. 106 Branch connections to stack.

The discharge of foul water from sanitary appliances is carried to the vertical discharge stack by branch pipes, as shown in Fig. 103. All branch discharge pipes should discharge into a discharge stack except those to appliances on the ground floor. Ground floor sinks, baths and wash basins may discharge to a gully, and WCs, bidets and urinals to a drain. A branch pipe from a ground floor WC should only discharge directly to a drain if the drop is less than 1.5 m, as shown in Fig. 104.

A branch pipe should not discharge into a stack lower than 450 mm above the invert of the tail of the bend at the foot of the stack for single dwellings up to three storeys, and 750 mm for buildings up to five storeys high.

The branch pipes from more than one ground floor appliance may discharge to an unvented stub stack, with the stub stack connected to a ventilated discharge stack or a drain, provided no branch is more than 2 m above the invert of the connection to the drain and branches from a closet more than 1.5 m from the crown of the closet trap, as shown in Fig. 105.

Branch pipes from waste water fittings such as sinks, baths and basins on the ground floor should discharge to a gully between the grating and the top level of the water seal. To avoid cross flow, small similar-sized connections not directly opposite should be offset by 110 mm on a 100 mm stack and 250 mm on a 150 mm stack as shown in Fig. 106. A waste water branch should not enter the stack within 200 mm below a WC connection as illustrated in Fig. 106.

Pipes serving single appliances should be at least the same diameter as the appliance trap and should be the diameter shown in Table 6 if the trap serves more than one appliance and is unventilated.

Bends in branch pipes, which should be avoided if possible, should have a radius as large as possible and a centre line radius of at least 75 mm for pipes of 65 mm or less in diameter. Junctions on branch pipes should be made with a sweep of 25 mm radius or at an angle of 45°, and connections to the stack of branch pipes of 75 mm diameter or more should be made with a sweep of 50 mm minimum radius or 45°.

Table 6 Common branch discharge pipes (unvented).

Appliance	Max. number to be connected	Max. length of branch (m)	Min. size of pipe (mm)	Gradient limits (fall per m)	
				min (mm)	max (mm)
WCs	8	15	100	9	90
urinals: bowls	5	*	50	18	90
stalls	7	*	65	18	90
washbasins	4	4 (no bends)	50	18	45

*No limitation as regards venting but should be as short as possible.

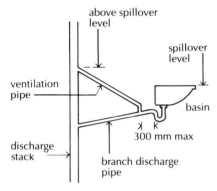

Fig. 107 Branch ventilation pipe.

It is not necessary to provide ventilation to branch pipes whose length and slope is limited to the figures given in Fig. 103, or to the common branch discharge pipes set out in Table 6. Where the length or slope is greater than these limits, the branch pipes should be ventilated by a branch ventilation pipe to external air, to a discharge stack or to a ventilating stack where the number of ventilating pipes and their distance to a discharge stack are large.

Branch ventilating pipes should be connected to the discharge pipe within 300 mm of the trap and should not connect to the stack below the spillover level of the highest appliance served, as illustrated in Fig. 107. Branch ventilating pipes to branch pipes serving one appliance should be 25 diameter or where the branch is longer than 15 m or has more than five bends, 32 in diameter.

Discharge pipes and stacks

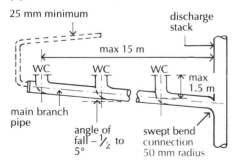

Fig. 108 Range of WCs branch discharge pipe.

Discharge pipes from ranges of WCs are usually 100 mm. The discharge pipes do not run full and there is little likelihood of self-siphonage of traps to appliances and, therefore, no need for venting. Ranges of eight or more WCs may be connected to a common discharge pipe. The shape and length of the common discharge pipe is not critical. The connection to the stack should be through a swept bend and the connections of the WCs to the common discharge pipe should likewise be through a swept bend as illustrated in Fig. 108. Where there are more than eight WCs in a range or more than two bends in the branch discharge pipe, a ventilating pipe may be necessary.

A range of up to four basins may be connected to a 50 mm common discharge pipe without venting, as illustrated in Fig. 109, and up to five with one 25 mm vent to the highest point of the discharge pipe. With ranges of more than five basins it is necessary to vent the trap to each appliance to limit pressure fluctuations that might otherwise

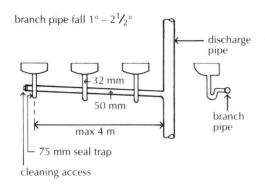

Fig. 109 Range of up to four wash basins with P traps.

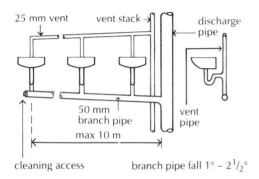

Fig. 110 Range of up to 10 wash basins with either P or S traps.

cause self-siphonage of the water seal to traps, as illustrated in Fig. 110. The vent stack and pipes are connected to the discharge pipes to each appliance and the vent stack may be run to outside air independent of the discharge stack.

The discharge pipes from basins fitted with spray taps do not run full and there is, therefore, no likelihood of self-siphonage. Ranges of up to eight basins fitted with spray taps do not require ventilation of traps.

The discharge pipe from urinals does not run full and there is no need for ventilation of traps. Discharge pipes to urinals should be as short as possible to minimise build up of deposits.

Ranges of WCs and basins and urinals on several floors may discharge to a common single stack without vent pipes when the estimated frequency of use of appliances and the consequent discharge loading of the pipe system is unlikely to cause gross pressure fluctuations.

Where the discharge loading of the stack is likely to cause pressure fluctuations sufficient to cause induced siphonage or back pressure, a vent stack connected to the discharge stack at each floor is used to limit pressure fluctuations. Where the traps of ranges of wash basins have to be vented to avoid self-siphonage, the vent stack may be run to outside air independent of the discharge.

Where both the discharge loading of the stack and the number of a range of wash basins require venting, the vent stack is connected both to the wash basin traps and the discharge stack as in the ventilated system. Fig. 99 illustrates these arrangements. Which one is used will depend on the estimated frequency of use, assumed discharge loading and the size of the discharge stack.

Alternatively, the vent stack may be connected to air independently of the discharge stack, as in the modified single stack system where the vent pipe system is used to limit pressure fluctuations in the discharge pipes to basins. Which system is used will depend on the estimated frequency of use, assumed discharge loading and the size of the discharge stack.

Discharge stacks should not have offsets in any part carrying foul water, should be run inside a building if it has more than three storeys and should discharge to a drain through a bend with as large a radius as possible and not less than 200 mm at the centre line, as illustrated in Fig. 103.

Discharge stacks should be ventilated to prevent water seals in traps being drawn by pressure that can develop in the system, by being continued up to outside air at least 900 mm above any opening in the building within 3 m and finished with a cage or cover that does not restrict flow of air.

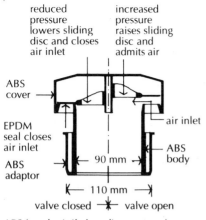

reduced pressure lowers sliding disc and closes air inlet

increased pressure raises sliding disc and admits air

ABS cover

EPDM seal closes air inlet

ABS adaptor

air inlet

ABS body

90 mm

110 mm

valve closed —✳— valve open

ABS (acrylonitrile butadiene styrene) air admittance valve for use with adaptor, reducer and ring seal coupling to uPVC pipes

Fig. 111 Air admittance valve.

A discharge stack may terminate inside a building if it is fitted with an air admittance valve (Fig. 111) which has a British Board of Agreement Certificate. An air admittance valve operates through a disc which rises, due to increased pressure in the discharge stack, to release the air pressure and falls when the pressure is reduced to normal air pressure. An EPDM (ethylene-propylene diene monomer) seal serves to close the air inlet.

The dry part of a discharge stack above the highest branch, which serves only for ventilation, may be reduced to at least 75 mm diameter on one and two-storey houses.

The size of the discharge stack is determined by the anticipated flow from all the fittings discharging into it. Rodding points should be provided in the stack to give access to any length of pipe that cannot be reached from another part of the system.

PIPE MATERIALS

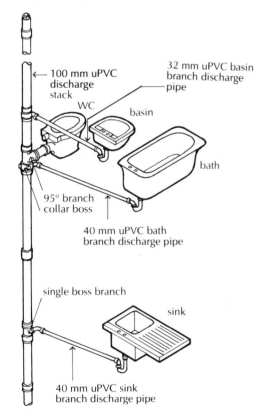

100 mm uPVC discharge stack

32 mm uPVC basin branch discharge pipe

WC

basin

bath

95° branch collar boss

40 mm uPVC bath branch discharge pipe

single boss branch

sink

40 mm uPVC sink branch discharge pipe

Fig. 112 uPVC discharge stack and branches.

The materials used for discharge pipework above ground are cast iron, copper and plastics. uPVC (unplasticised polyvinyl chloride) is the most commonly used material for discharge stack and branch pipe systems, because of its low cost, ease of cutting, speedily made joints and the range of fittings available. This smooth surface material is usually finished in black or grey, which requires no protective coating. The pipework is secured with loose brackets that are nailed or screwed to plugs in walls.

A variety of fittings is manufactured to suit the various branch waste connections for single stack systems of sanitary pipework. The plastic pipework is usually jointed by means of an elastomeric ring seal joint. The synthetic rubber ring forms an effective seal as it is compressed between the spigot and socket ends of the pipe as the spigot end is pushed into the socket.

Connections of uPVC branch pipes to outlets of appliances is made with a rubber compression ring that is hand tightened by a nut to a copper liner. Fig. 112 is an illustration of a uPVC discharge stack to a house.

Cast iron pipes

Cast iron pipes are used for discharge pipework where the strength and durability of the material and the wide range of fittings available justify the comparatively high initial cost. Where cast iron is used it is usual to run the discharge stack, WC discharge pipes and main branch pipes to ranges of fittings in cast iron, and discharge pipes from waste appliances in copper or plastic tube. In this way the strength and speed of fixing of cast iron is combined with the compact joints and ease of manipulation of copper or plastic pipe.

The pipes and fittings are sand cast or spun with socket and spigot ends, as illustrated in Fig. 113, and are given a protective coating of tar or bitumen. A range of socket and spigot fittings with or without bolted access doors is provided. Joints are made with molten lead which is caulked (rammed) into the joint that has been sealed with a gaskin of hemp, or with lead wool, caulked fibre or a rubber seal ring. Pipes and fittings are usually fixed by nailing through cast on ears to plugs in walls and floors.

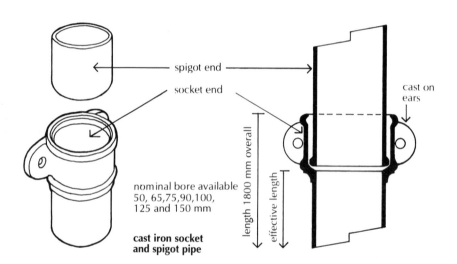

spigot end

socket end

cast on ears

length 1800 mm overall

effective length

nominal bore available 50, 65, 75, 90, 100, 125 and 150 mm

cast iron socket and spigot pipe

Fig. 113 Cast iron pipes.

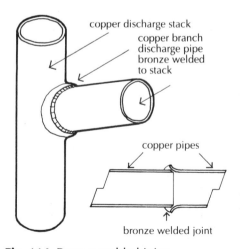

copper discharge stack

copper branch discharge pipe bronze welded to stack

copper pipes

bronze welded joint

Fig. 114 Bronze welded joint.

Copper pipe (tubulars) with capillary or compression joints is used for discharge pipes to waste appliances because of the compact joints in this material, its ease of handling, and particularly for the facility of making bends in pipework run from appliances through walls to connect to discharge stacks.

With extensive discharge stack and branch pipework run internally in ducts, prefabricated systems of copper pipework may be used. The pipework is jointed with bronze welded joints (Fig. 114), so that complete sections of pipework may be prepared off site ready for fixing, to minimise site work. Here the high initial cost of pre-fabricated pipework is justified by appreciable reduction in on-site labour working in cramped conditions and also the durability of the pipework.

TESTING

Soundness test (air test)

The accepted method of testing the soundness of discharge stacks and pipes above ground is the air test. A sound pipe system will contain air under pressure for a few minutes as an indication of its capacity to contain the flow of liquid in conditions normal to a discharge pipe system.

The air test is carried out to the whole discharge pipework above ground in one operation or, where the pipework is extensive, in two or more operations. The traps of all sanitary appliances are filled with water and the open ends of pipes are sealed with expanding drain plugs or bag stoppers. Air is pumped into the pipework through the WC pan trap and the air pressure is measured in a U tube water gauge or manometer. A pressure equal to 38 mm water gauge should be maintained for at least three minutes if the pipework is sound. Fig. 115 illustrates the equipment used for the air test.

If the air pressure is not maintained for three minutes, leaks may be traced by spreading a soap solution around joints, with the pipework under air pressure; bubbles in the soap solution will indicate leaks or, alternatively, smoke is pumped into the pipework from a smoke

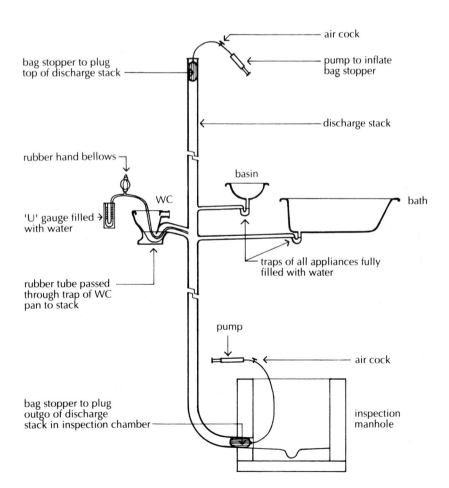

Fig. 115 Air test for sealed discharge stack and branches.

machine and the escape of smoke will indicate leaks. Leaking joints are made good and the air test applied to test for soundness.

In use test

To test the performance of a discharge pipe system in use, a group or groups of appliances are discharged simultaneously to cause conditions most likely to produce maximum pressure fluctuations. In buildings with up to nine appliances of each kind to a stack, the top-floor sink and wash basin are filled to overflowing and the plugs pulled simultaneous to a normal discharge of the top-floor WC. After this test a minimum of 25 mm water seal should be retained in every trap. With more than nine appliances of each kind to a stack, two or more WCs, basins and sinks are discharged simultaneously on the top floors for the performance in use test.

The discharge from fittings to the top of a stack provides conditions most likely to cause pressure fluctuations sufficient to induce siphonage and back pressure and loss of water seal. In the performance in use test, the discharge from baths, showers and urinals is ignored as their use does not generally add significantly to peak flow conditions.

PIPE DUCTS

For some years it was a requirement that discharge stacks be run internally to avoid the possibility of water in waste branches freezing and causing blockages. In the temperate climate of England it is unlikely that this will occur as waste branches rarely run full bore and running water does not readily freeze. The water from a slowly dripping tap may however gradually freeze and cause a blockage.

Because of the need to run internal discharge stacks inside pipe casing or ducts, for appearance sake and to avoid the inconvenience of unblocking drains internally, the requirement has by and large been abandoned, particularly for small buildings where there is access to discharge stacks run externally.

In large buildings all discharge pipework, together with hot and cold water services, are run internally in ducts for ease of access.

Where there are internal bathrooms and WCs to economise in pipework and to avoid over large ducts it is necessary to group sanitary appliances. Some compact groups of sanitary appliances and ducts and pipework are illustrated in Fig. 116, which illustrates ducts to bathroom with WC and a separate bathroom and WC.

Where there are two or more bathrooms and WCs on each floor of a multi-storey building, they will be grouped around a common duct: side by side where there are two, and back to back to a single duct where there are four.

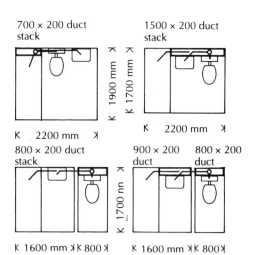

Fig. 116 Pipe ducts.

Ventilation of internal WCs and bathrooms

Bathrooms and WCs are often sited internally in modern buildings, such as flats, so that external walls may be best used for the windows of other rooms. It is therefore necessary to provide means of extract

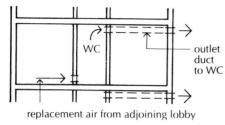

Horizontal outlet duct

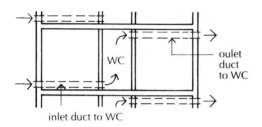

Horizontal inlet and outlet ducts

Fig. 117 Ducts.

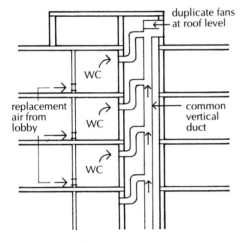

Fig. 118 Mechanical extract system.

ventilation to internal bathrooms and WCs, to rapidly dilute pollutants and moisture vapour in the air by air changes.

Extract ventilation may be provided either by mechanical extract of air or by passive stack ventilation. Passive stack ventilation is a ventilation system that uses ducts from the ceilings of rooms to terminals on the roof to operate through the natural stack effect of heated air rising, as in a chimney stack.

Mechanical ventilation is effected by an electrically operated extract fan designed to evacuate air rapidly through a duct to outside air. The Building Regulations recommend mechanical extract ventilation of 60 litres a second for bathrooms and sanitary accommodation located internally. The extract fan should be controlled by the operation of the light switch to the room and have a 15 minutes overrun after the light is turned off. To replace evacuated air there should be a 10 mm gap under the door to bathrooms and sanitary accommodation, through which replacement air can enter.

Air evacuated from internal bathrooms and sanitary accommodation is conducted through metal or plastic ducts run at high level or in hollow floors to outside air. Replacement air may enter at low level from an adjoining ventilated room either under a door or through a ventilation grille, as illustrated in Fig. 117. Where adjoining rooms or spaces afford poor ventilation for replacement air, it is necessary to provide ducts to supply replacement air from outside at low level, as illustrated in Fig. 117.

Mechanical ventilation through a vertical duct is used to provide steady air changes independent of wind direction and pressure. Air is drawn through a common vertical duct to the outside air with branches to individual internal rooms as illustrated in Fig. 118. A main fan and duplicate standby fan at roof level evacuate air through the vertical duct and shunt ducts from rooms. Replacement air enters the internal rooms from adjoining lobbies.

4: Foul Drainage

PIPE MATERIALS

Short, cylindrical, vitrified clay pipes have for centuries been used for drains underground. The joints between the pipes were made with puddled clay, either packed around the butt ends of adjacent pipes, packed into loose clay collars joining pipe lengths or packed into the socket end of one pipe around the spigot end of the next.

Puddled clay is plastic clay either in its natural wet state or to which water is added to make it plastic, so that it can be moulded around or packed tightly into collars or sockets of pipes. A gaskin of hemp, usually tarred or coated with tallow, was first rammed into the collars or sockets to align the pipes and prevent the puddled clay entering the pipeline. The clay pipeline was laid on the bed of the trench, then back filled with the excavated material.

Clay pipes with puddled clay joints

Well burnt (vitrified) clay or stoneware pipes were inert to sewage and impermeable to the intrusion of ground water and the puddled clay joints might remain watertight for many years, particularly in damp soil conditions. The plasticity of the clay joint would take up slight movements in the pipeline due to settlement, elongation or contraction of the pipeline, and slight displacement of the pipeline caused by backfilling the pipe trench and pressure on the ground from above.

The small movements that the puddled clay joints could accommodate might allow some seepage of sewage water through the joints to the pipeline, or intrusion of ground water into the pipe. The pipeline was buried underground, and unless considerable leaks from, or blockages of, the pipeline demanded attention, it remained so.

Up to the middle of the nineteenth century, standards of hygiene were appreciably poorer than those of today. The common practice in towns was to drain sewage into cesspits (pits dug into the ground, near to and sometimes under buildings) which retained solids and released liquids into open ditches and rivers, causing an all-pervading odour.

Following serious epidemics of typhus in the early nineteenth century, due to gross pollution of drinking water by sewage, there was a demand for drastic and rapid improvement in hygiene. During the latter part of the nineteenth century there was great activity in the building of new enclosed sewers to replace cesspits and an overall improvement in drain laying and maintenance.

Cement joints

By the beginning of the twentieth century the new wonder material, Portland cement, was being manufactured in quantity and the cement joint for drains, and later the concrete bed, were adopted as a 'cure-all' for all time, for blocked or leaking drains. The notion was to make a dense rigid joint of cement and sand between the brittle (rigid) clay pipes with a view to a rigid pipeline that would remain tight to seepage from inside and infiltration of ground water from without. To make doubly certain, the rigidly jointed clay pipeline that was initially laid on the bed of a drain trench was later laid on a rigid concrete base laid in the trench bottom. This combination of rigidly jointed clay pipes on a solid concrete bed was accepted as sound drain-laying practice from the beginning to the middle of the twentieth century.

A vitrified cylindrical drain pipe is brittle or rigid and will, under load, crack. The great advantage of the clay pipeline with puddled clay joints was that although the short lengths of pipe could not in themselves accommodate movement, the many plastic puddled clay joints could and did so without excessive seepage from or infiltration of ground water into the joints.

The movements that a pipeline underground may suffer are ground settlement, movement due to gain or loss of water in clay soils, disturbances during backfilling of the pipetrench and elongation or contraction due to temperature or moisture changes in the pipes and joints. After World War II (1939–45) the Building Research Station began an investigation of drain failures that culminated in Digests 124 and 125 (first series). The principal failures reported were due to blockage of the drain or excessive seepage from the drain requiring the attention of a builder.

At that time the two materials most in use for drain pipes were the traditional salt-glazed clay and the recently introduced pitch fibre. The investigations established that drain failures were due in clay pipelines to misalignment of the pipes, brittle fracture of the pipes or fracture of the rigid cement joint; and in pitch fibre lines they were due to flattening of the pipe or fracture of the joint coupling. These failures were caused by earth movement under or around the pipeline, or load stress on the pipeline from backfilling the trench, or surcharge loads on the ground above the trench, these causes often being made worse by supports placed under pipes during laying to facilitate alignment. Other causes of failure were temperature and moisture changes in the pipeline, often after laying and before backfilling the trench, and also damage to pipes during handling.

Rigid and flexible pipes

Clay pipes generally failed by brittle fracture either across or along the pipe or around sockets, whereas pitch fibre pipes generally failed by being flattened without fracture. The different behaviour of clay and pitch fibre pipes under load prompted the current classification of pipes as rigid and flexible.

Rigid pipes

Rigid pipes are those that fail by brittle fracture before they suffer appreciable deformation, and these include clay, concrete, cement and cast iron. Flexible pipes are those that suffer appreciable deformation before they fracture, and these include pitch fibre and uPVC.

Many of the failures of clay drainlines were a consequence of the rigidity of the pipes, the cement joints and the concrete bed in use at the time of the investigation. The rigidity of the pipeline and its bed were incapable of accommodating, without fracture, the soil movements and load stresses that a pipeline may suffer. From this understanding of failures there developed the discontinuous concrete or granular bed, and the flexible jointing system of clay drainlines. Thus practice had gone full circle from the flexible puddled clay joint, through rigid cement joints and bed, back to the flexible joint and granular bed of today. The deformation of flexible and rigid pipes has been controlled by the use of a granular bed, and limitations of load by the design of the trench and its backfilling.

DRAINAGE LAYOUT

The layout of foul drains depends on whether foul water and rainwater are discharged to a common drain system or to separate drain systems, which in turn depends on whether there is one sewer carrying both foul and rainwater or separate sewers for foul and rainwater.

Fig. 119 is an illustration of combined and separate drainage systems to a small two-floor house. As the drains for foul water and rainwater will generally run across each other at some point, it is necessary to adjust the level or gradient (slope) of the drains to accommodate this.

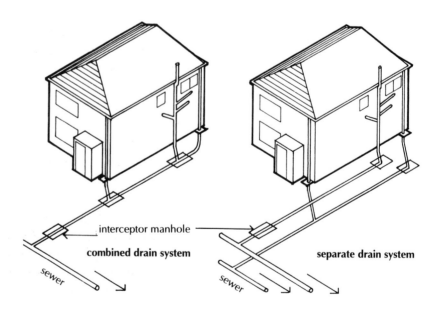

Fig. 119 Combined and separate drain systems.

Combined drains

In many of the older urban areas of England there is one sewer that takes the discharge of both foul water and rainwater from roofs and paved areas. Foul water is the discharge from WCs, bidets, baths, basins and sinks, and rainwater is the discharge of the run-off of rainwater from roofs and paved areas.

When the single, combined sewers of the older urban areas were laid out there was, at best, only rudimentary filtration and discharge from sewers to sea, river or inland soakaways. With later, more stringent controls for the purification of foul water, to minimise contamination of drinking water it became convenient and economic to separate foul and rainwater discharges to reduce the volume of water discharged to foul water sewers and the necessary size of water purification plants.

Separate drains

Outside the older urban areas it is usual for foul water and rainwater to drain to separate drainage and sewer discharge systems. It is general practice, therefore, to have combined drainage and sewer systems in the older urban areas and separate foul and rainwater systems in most other areas.

A combined drainage system which carries both foul and rainwater has to be ventilated throughout to conduct foul air discharge to the open air. In this system it is necessary, therefore, to fit trapped gullies with a water seal, to collect rainwater from roofs and paved areas. With separate systems of foul and rainwater drains it is necessary to fit trapped gullies only to the discharge of foul water fittings, where the discharge pipe does not serve as a ventilation pipe.

For economy in the use of labour and materials, the layout of a drain system should be kept simple. Fittings that discharge foul water should be grouped together on each floor to economise in water service pipe runs, discharge pipe branches should run to a common waste stack and groups of fittings on each floor should be positioned one over the other to avoid wasteful runs of pipework. Single fittings, such as basins or sinks fitted distant from other fittings, involve uneconomic and often unsightly lengths of water and discharge pipes.

Rainwater pipes from roof gutters and gullies to collect water from paved areas should be positioned to economise in and simplify drain runs. Wherever practicable, changes of direction and gradient should be as few and as easy as possible to minimise access points necessary to clear blockages.

Foul water drainage systems should be ventilated by a flow of air, with a ventilating pipe to the head of each main drain and any branch drain that is more than 6 m long serving a single appliance or 12 m long serving a group of appliances. Ventilated discharge pipes such as discharge stacks, discharging directly to the drain, are commonly used for ventilation of drains.

Drain runs should be laid in straight lines wherever possible to encourage the free flow of discharge water by gravity, with gentle curves in drain runs only where straight runs are not practicable. Bends in drain runs should be limited to positions close to or inside inspection chambers and to the foot of discharge pipes and should have as large a radius as practicable.

Where drain runs are near to or under a building, precautions should be taken to accommodate the effects of settlement without damage to the drain.

DRAIN PIPES

Clay pipes – materials, manufacture, sizes

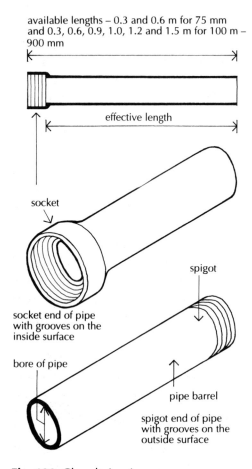

available lengths – 0.3 and 0.6 m for 75 mm and 0.3, 0.6, 0.9, 1.0, 1.2 and 1.5 m for 100 m – 900 mm

effective length

socket

socket end of pipe with grooves on the inside surface

bore of pipe

spigot

pipe barrel

spigot end of pipe with grooves on the outside surface

Fig. 120 Clay drain pipes.

Up to the middle of the nineteenth century clay pipes were made from local clays and fired in primitive kilns. Depending on the type of clay used, the skill in moulding and the control of the firing, pipes varied from well-formed dense pipes to soft, porous, badly-shaped pipes.

The increase in urban population that followed the Industrial Revolution, and the demand from the middle of the nineteenth century for improvement in hygiene, brought increases in the production and quality of clay pipes. About this time it became common practice to use salt-glazed clay drain pipes coated inside and out with a dense impermeable glaze. The fired-on salt glaze rendered both dense and porous pipes impermeable to both infiltration and exfiltration of water.

More recently, through quality control, preparation, moulding, drying and firing, clay pipes are produced which are sufficiently dense and impermeable in themselves so that they no longer require salt glaze against exfiltration and infiltration of water. Old habits die hard, however, and in spite of improved products and techniques the clay pipe manufacturers produce vitrified clay pipes glazed inside only or both inside and outside, to satisfy the demands of traditionalists who claim without justification that a glaze encourages flow in a pipe.

Today the majority of clay pipes are mass produced in a few highly automated plants. The selected clays are ground to a fine powder and just sufficient water is added for moulding. Pipes are formed by high-pressure extrusion – socketed pipes individually and plain pipes continuously – the cylinder of formed clay being cut to length. Straightforward fittings are formed by extrusion and the parts of junctions are extruded and then cut and joined by hand.

The moulded clay pipes and fittings are then dried in kilns to encourage regular, uniform loss of water to avoid loss of shape. Pipes and fittings are automatically fed through and fired in continuous tunnel kilns. Pipes and fittings which are to be glazed have a ceramic

Fig. 121 Clay drain pipe fittings.

Jointing

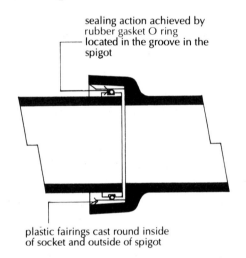

sealing action achieved by rubber gasket O ring located in the groove in the spigot

plastic fairings cast round inside of socket and outside of spigot

Fig. 122 Section at junction of socket and spigot clay pipes.

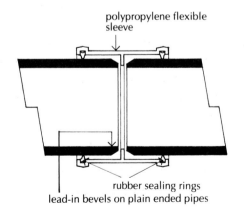

polypropylene flexible sleeve

rubber sealing rings
lead-in bevels on plain ended pipes

Fig. 123 Section at junction of plain ended clay pipes showing flexible coupling.

glaze or slip sprayed on, or they are dipped in the slip before they enter the kiln.

The nominal bore (inside diameter) of clay pipes for drains is from 75 to 900 mm in increments of 25 mm between 75 and 250 mm, one increment of 50 mm to 300 mm and then increments of 75 mm from 300 to 900 mm, as illustrated in Fig. 120.

A wide range of more than 250 fittings is made for clay drains such as bends, junctions, channels and gullies, some of which are shown in Fig. 121.

The advantage of the clay pipe for drains is that the comparatively short length of pipe and the wide range of fittings are adaptable both to the straightforward and the more complex drain layouts and the pipes themselves are inert to all normal effluents.

It is now generally accepted that flexible joints should be used with rigid clay drain pipes so that the drainlines are flexible in the many flexible joints. Flexible joints will accommodate earth movements under, around or over the pipeline. The flexible joint can be made in all weather conditions and, once made and tested, the trench can be backfilled to protect the pipeline from damage. There are two types of flexible joint in use: the socket joint for socket and spigot pipes and the sleeve joint for plain-ended clay pipes. Typical joints are shown in Figs. 122 and 123.

These flexible joint seals are made from either natural rubber, chloroprene rubber, butyl rubber or styrene-butadiene rubber. The flexible socket joint is made with plastic fairings cast on the spigot and socket ends of pipe to provide a simple push fit joint. It suffers the disadvantages that the joint may be damaged in handling and cut pipe lengths present difficulties on site.

The flexible coupling joint is made with a close fitting, flexible plastic sleeve with rubber sealing rings. The ends of the plain-ended pipes are bevelled to assist in forcing pipe ends into the sleeve to make a close watertight seal. Care is required to make these flexible joints in fitting pipe ends together and into sleeves to form a tight fit without damaging the seals.

The traditional cement and sand, rigid joint is still, to some extent, used by traditionalists on the grounds that it has been in use for a long time and is comparatively simple to make. A gaskin of hemp is first rammed into the joint between the spigot end of one pipe and the socket end of the other, and a mix of cement and sand is then rammed in to complete the rigid joint. Providing there is no undue settlement this joint may remain reasonably sound and watertight for some considerable time.

Cast iron pipes – materials, manufacture, sizes

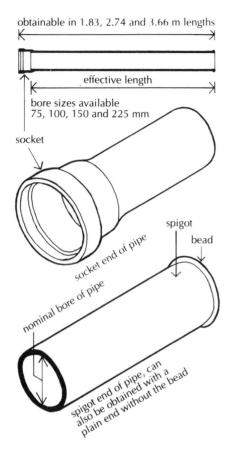

obtainable in 1.83, 2.74 and 3.66 m lengths

effective length

bore sizes available
75, 100, 150 and 225 mm

socket

socket end of pipe

nominal bore of pipe

spigot

bead

spigot end of pipe, can also be obtained with a plain end without the bead

Fig. 124 Cast iron drain pipes.

Jointing

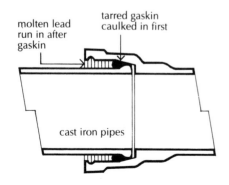

molten lead run in after gaskin

tarred gaskin caulked in first

cast iron pipes

Fig. 125 Rigid joint made with run pig lead.

For over 100 years cast iron pipes have been used for drains underground and for water and gas mains. A cast iron pipe is brittle or rigid and may suffer brittle fracture under heavy load or due to careless handling, but cast iron has better resistance to corrosion than either steel or wrought iron. In its molten state it runs freely into moulds, producing good, sharp, plain or intricate shapes. A cast iron pipe is stronger than the traditional clay pipe and has largely been used for drains under buildings and roads because of its superior strength. Cast iron pipes are coated inside and out with a hot dip solution of tar or bitumen to protect the iron from sewage and ground water.

Some years ago ductile iron pipes were first introduced. Molten grey iron (cast iron) is treated to change the micro-structure of the metal from the brittle graphite flake of cast iron to the ductile spheroidal graphite. The effect of this change is that the iron becomes ductile and can suffer deformation under load without fracture. The ductile iron pipe is thus a flexible pipe having high tensile strength, ductility and resistance to severe impact without fracture.

Cast iron pipes are used for drains because of their superior strength where there is unstable or made up ground, in shallow trenches, under buildings, for drains suspended under floors of buildings, in heavily waterlogged ground and where sewage is under pressure from pumping.

The traditional method of manufacturing cast iron pipes is to pour molten grey iron into vertically mounted sand moulds. More recently, cast iron pipes are made by spinning inside a mould. Molten grey iron is poured into a revolving water-cooled mould in which the molten metal forms on the inside of the mould by centrifugal force, producing a pipe of even thickness and smooth finish. Casting by this method is rapid and continuous. Ductile iron pipes are also made by the spin-moulding process. Cast iron and ductile iron pipes are made with socket and spigot ends or with plain ends, as shown in Fig. 124. All pipes are hot dip coated with either a bituminous or tar coating inside and out.

Pipes are manufactured with bore of 75, 100, 150 and 225 mm and in lengths of 1.83, 2.74 and 3.66 m, together with a wide range of fittings.

The traditional joint for socket and spigot cast iron pipes is the run lead or lead wool joint. A tarred jute gaskin is rammed (caulked) into the socket to align the pipes and prevent lead entering the pipeline. Molten lead is then run into the socket against an asbestos clip used to retain the molten lead joint, as shown in Fig. 125. This is a skilled laborious task that can only be carried out in dry weather, and

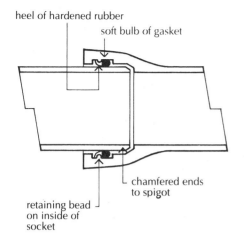

Fig. 126 Flexible push fit joint for cast iron.

requires room for working and clean conditions, all of which rarely combine in a drain trench. Lead wool – finely shredded lead – may be used instead of molten lead. The lead wool is hammered into the socket until it makes a watertight joint. This again is a laborious process.

A flexible push fit joint is commonly used today for cast iron and ductile iron pipes. A rubber gaskin, fitted inside the socket of pipes, comprises a heel of hardened rubber that aligns the pipes, and a bulb of softer rubber that makes the joint, as shown in Fig. 126. The pipe ends must be clean and are lubricated and joined by leverage from a crowbar for small pipes, or a fork tool for larger pipes. The joint is flexible and will accommodate longitudinal and axial movements. The joint is rapidly made in any conditions of weather, the pipeline may be tested right away and the trench can be backfilled to protect the pipeline.

The current recommended methods of laying rigid pipes on granular beds or on concrete, and the methods of protecting pipes from damage by settlement or from loads on the surface are described later in this chapter under the heading 'Drain laying'.

uPVC pipes – materials, sizes

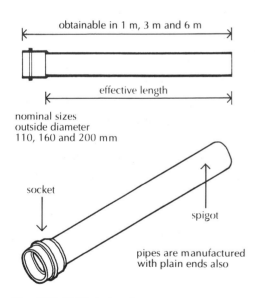

Fig. 127 PVC drain pipes.

uPVC pipes were first used for underground gas and water pressure pipes some 60 years ago. This pipe material was first used in the UK some 30 years ago and is now extensively used because of its ease of handling, cutting and jointing and the low cost. Polyvinyl chloride (PVC) is made by the electrolysis of coal and chalk to form carbide. Water is added to form acetylene. Hydrochloric acid is then added to form the monomer of vinyl chloride. The monomer is polymerised to a fine white powder of low density. To the PVC are added small quantities of lubricants, stabilisers and pigments. Heated uPVC material is extruded through a former and then 'frozen' by cooling to form a continuous pipe length.

The pipe is light in weight and flexible as it can to some extent deform under load without fracture. Deformation should be limited to an increase in horizontal diameter of not more than 5% to avoid blockages in pipelines or breaks to joints. Pipe sizes are described by the outside diameter of the pipe as 110, 160 and 200 mm, and lengths are 1.0, 3.0 and 6.0 m (Fig. 127).

Because of the comparatively long lengths in which this pipe material is made, with consequently few joints necessary, the material lends itself to assembly at ground level from where it can be lowered into narrow trenches. This is a distinct advantage on most building sites.

Jointing

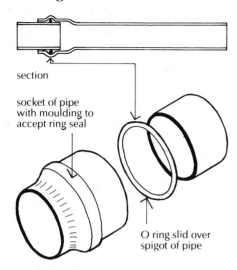

section

socket of pipe
with moulding to
accept ring seal

O ring slid over
spigot of pipe

Fig. 128 PVC socket and spigot push fit joint.

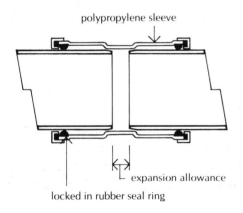

polypropylene sleeve

expansion allowance

locked in rubber seal ring

Fig. 129 PVC sleeve joint.

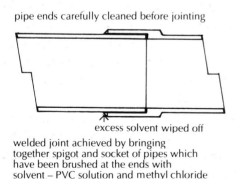

pipe ends carefully cleaned before jointing

excess solvent wiped off

welded joint achieved by bringing
together spigot and socket of pipes which
have been brushed at the ends with
solvent – PVC solution and methyl chloride

Fig. 130 PVC solvent welded joint.

Two jointing systems are used, as described here.

Socket and spigot push fit joints with rubber rings depend on a shaped socket end of pipe, as shown in Fig. 128, or a shaped coupling designed to fit around rubber sealing rings that fit between the socket or coupling and spigot end to make a watertight seal. Because the rubber ring seal has to be a tight fit between the pipes, it requires careful manipulation to fit the spigot end into the socket end of the pipe or coupling while making certain the rubber seal takes up the correct position in the joint.

Because of the considerable thermal expansion and contraction of this material it is necessary to provide for movement along the pipe lengths through the allowance between the spigot end of pipes and the shoulder on the socket end of the pipe or coupling, as illustrated in Fig. 128.

Several proprietary systems of push fit joints are available.

Sleeve joints of polypropylene with rubber sealing rings to plain-ended pipes depend on a separate plastic coupling sleeve that fits over the spigot ends of plain pipe lengths. Rubber seal rings in the ends of the sleeve compress on the spigot ends of the pipes being joined, as illustrated in Fig. 129. To provide a tight fit between the coupling sleeve and the pipe ends it is necessary to apply some force to push the pipe ends into the sleeve, for which cramps or jacks are available. The ends of the pipes being joined fit to shoulders in the sleeve to provide the expansion allowance illustrated in Fig. 129.

This sleeve jointing is generally favoured over the socket and spigot joint.

Solvent welded joints (Fig. 130) are available for short lengths of drain pipe. This type of joint is not used for long lengths of drain, as the rigid joint makes no allowance for expansion. The joint is made by bringing together the spigot and socket ends of the pipe after they have been brushed with a PVC solution and methyl chloride. The solvent dissolves the PVC, which fuses together after some time. A disadvantage of this joint is that it takes some time to fully harden.

uPVC drain pipes are commonly used with the proprietary drain systems such as the 'rodding eye' or the 'access bowl' drainage systems shown in Figs. 146 and 147. These systems comprise a package of plastic connections, clearing eyes and access bowls for inspection, to combine economy and speed of assembly and laying.

Because of the comparatively extensive range of fittings manufactured in clay, it is not uncommon for clay gullies, for example, to be used in conjunction with uPVC pipes.

Pitch-fibre pipes – materials, manufacture, sizes

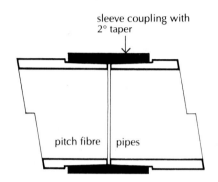

Fig. 131 2° taper coupling joint – long section through joint.

Jointing

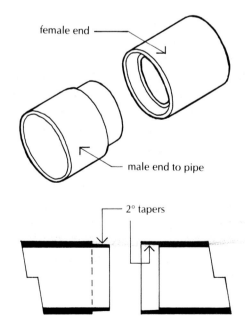

Fig. 132 Section through spigot and socket joint in pitch fibre pipes.

Pitch-fibre pipes have been successfully used on the Continent and America for many years. They were first manufactured and used in the UK from the middle of the twentieth century and were used for drains underground for both soil and surface water. The low cost, long length of pipe and ease of jointing, together with the flexibility of the material, recommended it as a drain pipe material.

Pitch fibre pipes are much less used for drainage than they were some years ago, principally because of the difficulty of making junctions and bends with the material and because of the very limited range of fittings. uPVC pipes have largely taken over as the principal flexible drain pipe material. Pitch fibre pipes are still used for long straight runs of surface water and land drainage and as conduits for small cable and pipe material.

The pipes are manufactured from a blend of wood cellulose products such as waste paper and other fibres. The fibres are mixed with water and woven into a felt of fibre-pipe size. The moulded fibre-felt pipe is dried and then immersed in hot pitch and later in water to give the pitch-fibre pipe a gloss finish. Pitch-fibre pipes consist of about 30% fibre and about 70% pitch. The pipes have a nominal bore of 50, 75, 100, 125, 150, 200 and 225 mm and are supplied in lengths of 1.7, 2.5 and 3 m.

The joints in use are as follows.

The taper coupling joint was the original method of jointing when pipes were first produced. The ends of the pipe are machined to a 2° shallow taper and a pitch fibre coupling with 2° tapers acts as a joint. The joint is made by driving pipe ends into the coupling to form a watertight joint (Fig. 131).

Where it is necessary to make a joint to plain ends of pipe, as for example where a non standard length of pipe is used, the cut plain ends of pipe are jointed by a polyproplene coupling sleeve. Rubber D section snap rings are fitted around the plain ends of pipes to be joined. The pipe ends are pushed into the sleeve and the rubber rings roll along the pipe ends until they snap close with the flat of the D section rings bearing on the pipe. This joint requires some skill and care in the making if it is to be successful.

A spigot and socket joint is made between the tapered machined spigot end of one pipe and the tapered machined socket end of the next, by pushing the ends together. The components of this joint are shown in Fig. 132. The small end overlap of this joint may not provide an entirely watertight joint, but that is of little consequence as this joint is used to align perforated pipes used for land drainage, and plain pipes used as conduits for cables.

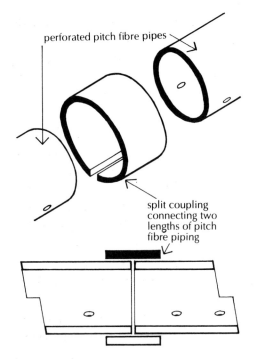

perforated pitch fibre pipes

split coupling connecting two lengths of pitch fibre piping

Fig. 133 Split coupling joint for pitch fibre pipes – long section through joint.

The split coupling joint (Fig. 133) is used for pipes perforated for land drainage and for pipes used as conduits. The advantage of this joint, which is used solely to align pipe lengths, is that it can be used to join non standard lengths.

Table 7 Recommended minimum gradients for foul drains.

Peak flow [litres/sec]	Pipe size [mm]	Minimum gradient [1:...]	Maximum capacity [litres/sec]
< 1	75	1:40	4.1
	100	1:40	9.2
> 1	75	1:80	2.8
	100	1:80*	6.0
	150	1:150†	15.0

* Minimum of 1 WC.
† Minimum of 5 WCs.

DRAIN LAYING

Drains laid underground should be of sufficient diameter to carry the anticipated flow, and should be laid to a regular fall or gradient to carry the foul water and its content to the outfall.

Pipe gradient or fall

Up to about 50 years ago traditional wisdom was to use a 4 in drain for up to four houses and a 6 in drain for more than four houses, the 4 in drain being laid at a uniform gradient of 1 : 40 and the 6 in drain at a gradient of 1 : 60. The gradient of the drain was related directly to the size of the pipe and bore little relation to the anticipated flow in the drain. The result was that drain pipes were often oversized and the gradient steeper than need be, resulting in excessive excavation.

Study of the flow load of drains in use has shown that the size of the drain pipe and its gradient should be related to the anticipated flow rate in litres per second, so that the discharge entering a drain will determine the necessary size and gradient of the drain. Where the sanitary appliances from a house or a few houses or flats discharge to a drain, the flow rate figure is so small as to require the smallest drain and it is not worth making a calculation of flow rate.

Drains are designed to collect and discharge foul and rainwater by the flow of water under gravity. Drains are, therefore, laid to a regular fall (slope) towards the sewer or outflow. The necessary least gradient or fall of a drain depends on the anticipated flow of water through it and the necessary size of drain to carry that flow. Table 7 gives recommended minimum gradients for foul drains.

Table 8 Discharge capacities of foul drains running 0.75 proportional depth.

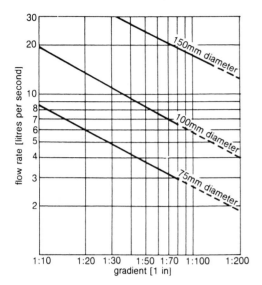

Approved Document H1 of the Building Regulations recommends that a drain carrying only waste water should have a diameter of at least 75 mm, and a drain carrying soil water at least 100 mm. The term waste water is generally used to include the discharge from baths, basins and sinks, and soil water includes the discharge from WCs.

Table 8, from Approved Document H1 of the Building Regulations, shows the relationship of flow rate to gradient for three pipe sizes with drains running three quarters of proportional depth. The rate of flow in drains with gradients as flat as 1:200 is given. Drains are not commonly run with gradients below 1:80 because the degree of accuracy necessary in setting out and laying drains required for shallow falls is beyond the skills of most building contractors.

Drains should be laid at a depth sufficient to provide cover for their protection and the excavation should be as narrow as practical for bedding and laying the drain lines. The greater the width of the trenches at the crown of the pipe, the greater the surcharge loads on the pipe. It is advantageous, therefore, to bed the drain in a narrow trench which may be increased in width above the level of the crown of the drain for ease of working. With modern excavating machinery, flexible joint pipelines may be assembled above ground and then lowered into and bedded in comparatively narrow trenches, so saving labour and cost in excavation and providing the best conditions for the least loads on the pipeline.

The depth of the cover to drain pipes depends on the depth at which connections are made to the drain, the gradient at which the pipes are laid and ground levels. Depth of cover is taken as the level of finished ground or paving above the top of a drain pipe. A minimum depth of cover is necessary to provide protection to the pipe against damage, and a maximum depth to avoid damage to the drain by the weight of the backfilling of the drain trenches. Minimum and maximum cover for rigid pipes are set out in Table 9.

Flexible pipes should have a minimum of 0.6 m of cover under fields and gardens and 0.9 m under roads.

Table 9 Limits of cover for standard strength rigid pipes in any width of trench.

Pipe bore	Bedding class	Fields and gardens		Light traffic roads		Heavy traffic roads	
		Min	Max	Min	Max	Min	Max
	D or N	0.4	4.2	0.7	4.1	0.7	3.7
100	F	0.3	5.8	0.5	5.8	0.5	5.5
	B	0.3	7.4	0.4	7.4	0.4	7.2
	D or N	0.6	2.7	1.1	2.5	–	–
150	F	0.6	3.9	0.7	3.8	0.7	3.3
	B	0.6	5.0	0.6	5.0	0.6	4.6

Bedding flexible pipes

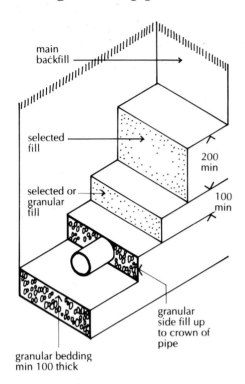

Fig. 134 Bedding for flexible pipes.

Bedding rigid pipes

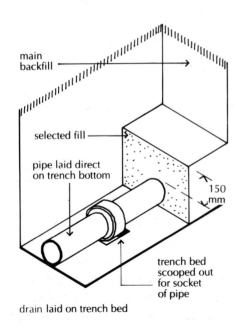

Fig. 135 Bedding for rigid pipes – Class D.

Flexible pipes should be laid in a narrow trench, on granular material such as clean, natural aggregate. This granular material is spread in the base of a trench that has been excavated and roughly levelled to the gradient or fall of the drains. The granular material is spread and finished to a thickness of 100 mm, to the drain gradient. Lengths of drain pipe are then lowered into the trench and set in position by scooping out the granular bed from under the collars of pipe ends, to set the pipe line in place in the centre of the trench. Further granular material is then spread and lightly packed each side of the pipe line, to support it against deformation under load, as illustrated in Fig. 134.

A layer of granular material or selected fill, taken from the excavated material, is then spread over the pipe line to a depth of 100 mm. A further layer of selected fill, free from stones, lumps of clay or other material larger than 40 mm, is spread in the trench to a thickness of 200 mm. The trench is then backfilled with excavated material up to ground level, and consolidated.

This drainlaying operation requires care and some skill to bed the drainline correctly, and further care in backfilling to avoid disturbing the drain. Where the bed of the trench is narrow, it may be necessary to cut the sides of the trench above the level of the bed of the drain, in the form of a V, sloping out from the centre for ease of access.

Where a drain trench is excavated in some cohesive soil such as clay for laying rigid pipelines of clayware, for example, it is possible to lay the pipes directly on the trench bottom which has been finished to the required pipe gradient by hand trimming by shovel. The pipe lengths are lowered into the trench and soil from the trench bottom is scooped out under each socket end of pipe so that the barrel of the pipes bears on the trench bottom and the collars keep the drainline in place. Fig. 135 illustrates this type of bedding.

For this operation to be successful the trench needs to be sufficiently wide for a man to stand in the trench astraddle the pipeline to scoop soil out from below the pipe collars. Once the pipeline is in place, a cover of selected fill from the excavation, free from large stones or lumps of clay, is spread in the trench to a depth of 150 mm above the crown of the pipes, and the trench is backfilled to surface level.

This bedding system is suited to the use of socket and spigot and clay pipes with one of the flexible joints that can be used in all weather conditions.

When the bed of drain trenches cannot be trimmed to the pipe gradient, a bed of granular material is spread in the bed of the trench and levelled to the pipe gradient to a thickness of 100 mm, as illustrated in Fig. 136. The rigid pipes are lowered into the trench and a layer of selected fill is carefully spread around the pipes and then filled up to a level 150 mm above the crown of the pipes. The trench is then back filled to surface level. This type of bedding is suited to plain clay pipes joined with flexible sleeves.

As an alternative, the granular bedding is spread 100 mm thick in the bed of the trench, the pipes are lowered into the trench and granular beddings is scooped out under the collar ends of pipes. The granular material is then spread around the pipes, up to half the outside diameter of pipe as shown in Fig. 137. Selected fill is then spread in the trench to a depth of 150 mm above the crown of pipes, and the trench is backfilled to the surface.

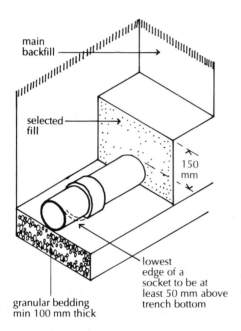

Fig. 136 Bedding for rigid pipes: granular bedding – Class N.

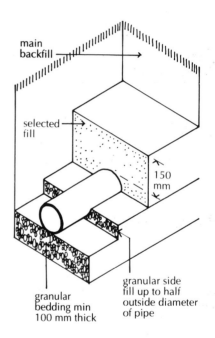

Fig. 137 Bedding for rigid pipes: Granular bedding – Class B.

This system of bedding has the advantage that the granular bedding each side of the pipes maintains them in position as the trench is filled.

It may well be that the most economical depth of flexible drain pipes below surface will provide less depth of cover than recommended. To avoid unnecessary excavation it is acceptable to lay flexible pipes with less cover when they are laid under fields or gardens. The pipes are laid on a bed of granular material 100 mm thick in the bed of the trench and then surrounded and covered with granular material up to a level of 75 mm above the crown of pipes, as illustrated in Fig. 138. Concrete paving slabs are then laid over the granular material, bearing on offsets in the trench walls. The trench is then backfilled up to surface level with selected fill from the excavation. The protection of concrete slabs and the granular material bed and surround of pipes will provide adequate protection against deformation of the pipes.

Rigid drain pipes that are laid at a depth that provides less than the recommended cover below surface, should be provided with protection from damage by an encasement of concrete, with flexible movement joints at each socket or sleeve joint of pipeline.

The drain trench is filled with concrete, to a depth of 100 mm, in sections along the length of the trench equal to pipe length. Between each section a 13 mm thick compressible board, holed for pipes, is set to the width of the trench. Before the concrete is hard it is scooped out for socket ends of pipe against one side of the board. A pipe is set in position up to the board and the next section of concrete is laid up to another flexible board in the trench. The concrete is scooped out for the next pipe, which is pushed into the flexible joint of the first pipe, and so on along the length of the trench, as illustrated in Fig. 139.

Once the drain line is laid it is tested and then concrete encasement is spread around the pipes, between the flexible boards, to provide a cover of 100 mm all round the drain. The trench is then backfilled to surface.

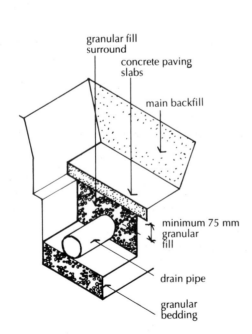

Fig. 138 Protection for flexible pipes under fields and gardens.

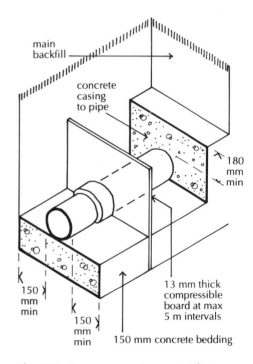

Fig. 139 Concrete casing to rigid pipes.

This complicated and expensive method of laying and protecting rigid drains is only used where no other method of laying would be possible.

Drains under buildings

When a drain is laid under a building and the crown of the pipes is at all points 300 mm or more below the underside of the ground floor slab, the drains should have flexible joints and be surrounded by granular material at least 100 mm thick all round the pipes. Where the crown of the drain is within 300 mm of the underside of the ground floor slab, it should be encased in concrete as an integral part of the slab.

Where a drain line is laid under a building there is a possibility that slight settlement of the wall might fracture the drain where it is laid to run through the wall. Drains under buildings are not uncommon in towns and cities where drains from the rear of buildings run under the buildings to sewers in the road.

There are two methods of avoiding the possibility of damage to drains running through walls.

The first method is by providing a minimum 50 mm clearance between the wall and the drain. The wall is built with a small lintel over the opening in the wall through which the pipe is to run, to provide at least 50 mm of clearance all round the drain. Where the drain is laid to run through the wall, two rigid sheets, holed for the pipe, are fitted to the pipe and secured in place by screws and plugs to the wall, as illustrated in Fig. 140, to exclude vermin or fill.

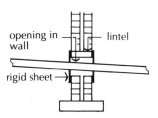

Fig. 140 Drain pipe through opening in wall.

The disadvantage of this system is that it is difficult to provide a close fit of a rigid board to a pipe and to brickwork and to make a watertight joint between the external board and the wall.

The other method of providing protection to a drain run through a wall, is to provide for any slight settlement of the wall by means of rocker pipes. A short length of pipe is built into the wall, projecting no more than 150 mm each side of the wall. A length of pipe at most 600 mm long is connected to each end of the built-in pipe and connected to it with flexible joints. These rocker pipes are likewise connected to the drain line with flexible joints so that any slight movement of the pipe built into the wall is accommodated by the flexible joints of the rocker pipes.

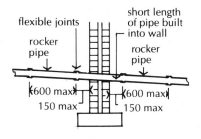

Fig. 141 Drain pipe built into wall.

This system, illustrated in Fig. 141, is best executed with plain clay pipes with flexible sleeve joints.

Drains close to buildings

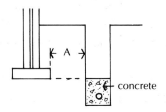

Fig. 142 Drains less than 1 m from foundations.

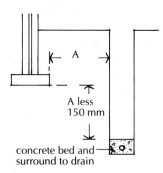

Fig. 143 Drains 1 m or more from foundations.

On narrow building sites, common to towns and cities, where there is only a narrow strip of land each side of a building, it may well be expedient to run a drain parallel to a flank wall of the building. To provide the necessary gradient or fall, the drain may be at or below the level of the wall foundation. To avoid the possibility of the loads on the wall foundation imposing undue pressure on the drain, it is often necessary to provide additional protection to the drain.

Where the drain trench bottom is less than 1 m from the foundation, indicated by the letter A in Fig. 142, the drain should be laid on a bed of concrete and then covered with concrete up to the level of the underside of the wall foundation.

A drain laid 1 m or more from the wall foundation, indicated by the letter 'A' in Fig. 143, and appreciably below the foundation of the adjacent wall, should be bedded on concrete and then covered with concrete up to a level equal to the distance from the wall of A less 150 mm.

This additional protection, which is expensive, may not be wholly satisfactory as it provides rigid encasement that makes no allowance for possible differential settlement along the length of the drain run, that might fracture the encasement and pipe run. It is plainly advisable to avoid drain runs close to walls where possible.

ACCESS POINTS TO DRAINS

There should be adequate access points to drains laid underground to provide means of clearing blockages by rodding through drains. Rodding is the operation of pushing flexible, sectional rods that can be screwed together, down drain lines to clear blockages. Drain rods may be of bamboo or other sufficiently flexible material capable of being bent to enter the drain line from an inspection chamber or manhole.

The four types of access point in use are:

(1) Rodding eyes, which are capped extensions of drain pipes
(2) Access fittings, which are small chambers in or as an extension of drain pipes with no open channel
(3) Inspection chambers, which are large chambers with an open channel but no working space at drain level
(4) Manholes, which are large chambers with an open channel and working space at drain level

The minimum dimensions for access fittings and chambers are given in Table 10.

Access points should be provided on long drain runs and at or near the head of each drain run, at a bend or change of gradient, a change of pipe size and at junctions where each drain run to the junction cannot be cleared from an access point.

The limits of the depth and minimum dimensions for access points are set out in Table 10 and the minimum spacing of access points in Table 11.

Table 10 Minimum dimensions for access fittings and chambers.

Type	Depth to invert [m]	Internal sizes		Cover sizes	
		length × width [mm × mm]	Circular [mm]	Length × width [mm × mm]	Circular [mm]
Rodding eye	—	As drain but min 100		—	
Access fitting small	0.6 or less	150 × 100	150	150 × 100	150
large		225 × 100	—	225 × 100	—
Inspection chamber	0.6 or less	—	190*	—	190*
	1.0 or less	450 × 450	450	450 × 450	450†
Manhole	1.5 or less	1200 × 750	1050	600 × 600	600
	over 1.5	1200 × 750	1200	600 × 600	600
	over 2.7	1200 × 840	1200	600 × 600	600
Shaft	over 2.7	900 × 840	900	600 × 600	600

* Drains up to 150 mm.
† For clayware or plastics may be reduced to 430 mm in order to provide support for cover and frame.

Table 11 Maximum spacing of access points in metres.

From	To	Access fitting		Junction	Inspection chamber	Manhole
		Small	Large			
Start of external drain*		12	12	—	22	45
Rodding eye		22	22	22	45	45
Access fitting small 150 diam		—	—	12	22	22
150 × 100		—	—	22	45	45
Inspection chamber		22	45	22	45	45
Manhole		22	45	45	45	90

* Connection from ground floor appliances or stack.

Inspection chamber – manhole

The traditional arrangement for inspecting, testing and clearing blockages in underground drains is the inspection chamber or manhole. This is a brick-lined pit at drain junctions and changes of direction or gradient in a drainline. The inspection chamber is located at those points where drain blockages are most likely to occur and from which blockages in drainlines can be cleared by rodding.

The inspection chamber provides access to inspect flow in the drain and, if necessary, means of testing drainlines. The traditional clay drain pipe was liable to blockages due to misalignment of the many joints or fracture of the pipes and their rigid cement joints, and there was therefore advantage in constructing inspection chambers at fairly frequent intervals when labour costs were low.

Today, inspection chambers are comparatively costly items and with the increased length of pipes available and flexible joints that closely align pipes and accommodate slight movement, it is possible to use fewer inspection chambers. The suppliers of uPVC drain pipes utilise systems of rodding points or access bowls instead of inspection chambers, for the purpose of clearing blockages. The rodding points cannot be used to inspect flow or for testing and it is usual to include one or more inspection chambers in this drain system.

The traditional inspection chamber or manhole – which is still much in use today where convenience outweighs cost and clay drain pipes are used – is a brick-lined pit at the junction of drain branches and at changes of direction and gradient, to facilitate inspection, testing and clearing obstructions. Their main purpose is access to clear blockages in any of the drain runs connecting inside the

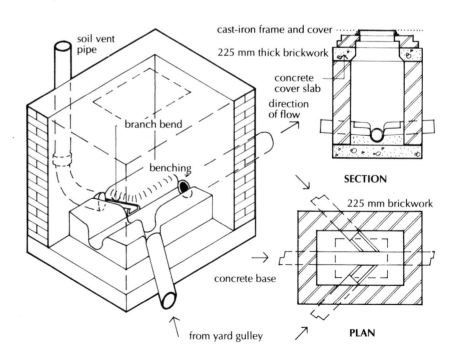

Fig. 144 Cut-away sectional view of inspection chamber.

chamber. An inspection chamber is a small, shallow chamber sufficient to clear blockages from above ground, and a manhole a deeper chamber large enough for a man to climb down to clear blockages.

An inspection chamber or manhole is formed on a 150 mm concrete bed, on which bricks walls are raised. In the bed of the chamber a half-round channel or invert takes the discharge from the branch drains, as illustrated in Fig. 144. The walls of the chamber may be of dense engineering bricks. If less dense bricks are used the chamber is lined with cement rendering to facilitate cleaning, and sometimes it is rendered outside to prevent the infiltration of groundwater. The chamber is completed with a cast iron cover and frame.

The word invert is used to describe the lowest level of the inside of a channel in an inspection chamber, or the lowest point of the inside of a drain pipe, and measurements to the invert of a drain are used to determine the gradient of that drain. In the bed of the chamber the three-quarter section branch drains discharge over the channel in the direction of flow, and fine concrete and cement rendering termed benching is formed around the branches to encourage flow in the direction of the fall of the drain, as shown in Fig. 144.

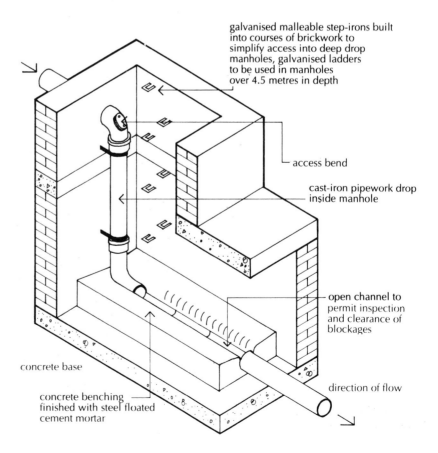

galvanised malleable step-irons built
into courses of brickwork to
simplify access into deep drop
manholes, galvanised ladders
to be used in manholes
over 4.5 metres in depth

access bend

cast-iron pipework drop
inside manhole

open channel to
permit inspection
and clearance of
blockages

direction of flow

concrete base

concrete benching
finished with steel floated
cement mortar

Fig. 145 Back drop manhole.

Backdrop manhole

Where a branch drain is to be connected to a main drain or a sewer at a lower level, it is often economical to construct a backdrop manhole to avoid deep excavation of a drainline. The backdrop chamber is constructed of brick on a concrete bed and the higher branch drain is connected to a vertical or drop drain that discharges to the channel in the backdrop chamber, as illustrated in Fig. 145 where the near side and end walls are omitted for illustration. This is a form of manhole adapted to suit the purpose. The drop drain is run in cast iron drains, supported by brackets screwed to plugs in the wall.

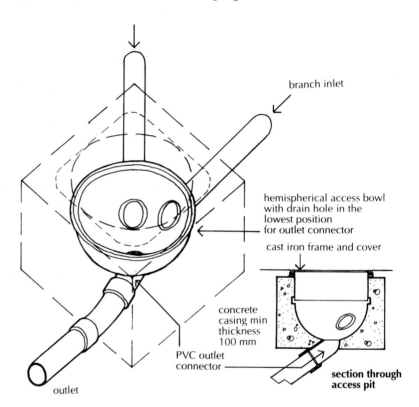

Fig. 146 Mascar access bowl drain system.

Access bowl drain system

As an alternative to the use of inspection chambers for access to drain junctions, where uPVC drains are used for small buildings such as houses, a system of PVC access bowls or chambers may be used.

The hemispherical access bowl (Fig. 146) is supplied with one hole for the outlet drain, and up to five additional holes can be cut on site to suit inlet branches. Purpose-made inlet and outlet PVC connectors are provided for solvent welding to the bowl for outlet and inlet connections to drains. One access bowl is sited close to the building for the discharge from the soil pipe and ground floor foul water and waste water fittings.

The access bowl is extended to ground level with PVC cylindrical extension sections, one of which is illustrated in Fig. 146. The access bowl is set on a concrete bed and after the drain runs have been connected it is surrounded with a minimum of 100 mm of concrete for protection and stability. A cast iron frame and cover is set on top of the access bowl at ground level.

These access bowls may be used at changes of direction in drain lines and at the boundary, before the connection to the sewer, for the purpose of rodding to clear blockages and inspection of flow. Access bowl access points, which are specifically designed for use with uPVC drain systems, can effect some cost saving compared to the traditional inspection chamber.

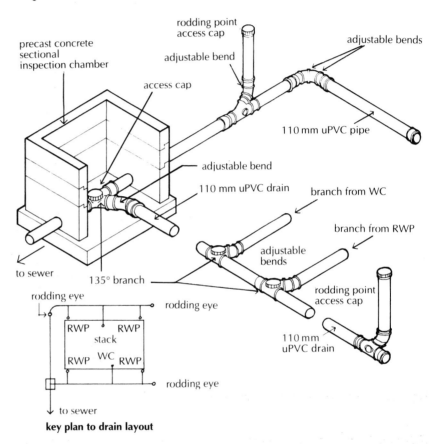

Fig. 147 Rodding point drain system.

key plan to drain layout

Rodding point drain system

This sealed underground drain system combines the advantages of the long lengths of pipe and simplicity of jointing uPVC drain lines, with rodding points and one or more inspection chambers for access for inspection, testing and clearing blockages. Fig. 147 shows a typical layout. The rodding points are used as a continuation of straight drain runs extended to ground level with an access cap, as illustrated in Fig. 147. It is possible to rod through each drain run to clear a blockage and this arrangement dispenses with the need for inspection chambers at all junctions and bends with appreciable saving in cost. The drain runs are laid on granular bedding and backfilled as previously described.

Access fittings with access caps may be provided at all drain junctions, as shown in Fig. 147. These access caps are provided for blockages at junctions where rodding along the drain is not effective in clearing the blockage.

The drain pipes of the rodding points, which extend to finished ground level, should be protected inside a length of clay drain pipe of larger diameter than the rodding drain, set on a bed of concrete under the drain line with concrete casing around the clay pipe liner. As an alternative, precast concrete, sectional circular ring liners, 150 mm internal diameter, may be set on a concrete bed and extended to ground level as protection for the rodding point.

The access cap to the junction inside the inspection chamber at the connection to the sewer, may be used to check flow in the drain and for rodding if need be.

The short branch drain connections to rainwater gullies, ground floor WC and soil pipe can be cleared by ferrets or other flexible clearing equipment. Access caps to junctions are often covered by backfilling and can be exposed by excavation for access in the rare event of a stubborn blockage.

Soil stack pipe to connection drains

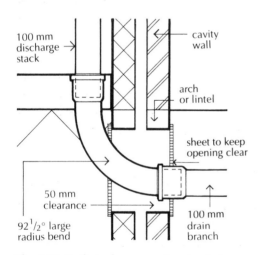

Fig. 148 Soil stack connection to drain.

In the single stack system, a single discharge stack pipe serves to carry both soil water and waste water discharges directly down to the drains, so that the soil pipe or stack may serve as a ventilation stack pipe to the drains. In this system the water trap seals to each of the sanitary appliances serve as a barrier to drain smells that might otherwise enter the building.

At the base of the soil pipe is a large radius or easy sweep bend to facilitate clearing blockages. In Fig. 148 the soil pipe is run inside the building – common practice in tall buildings. Whether the soil pipe is run inside or outside the building there should be a large radius bend at its connection to an inspection chamber.

Where the drain connection to a soil or waste pipe passes through the wall of the building, there should be at least 50 mm of clearance all round the drain to allow for any settlement of the wall that might otherwise fracture the pipe if it were built into the wall. The opening in the wall around the pipe is supported by small lintels or brick arches and covered with rigid sheet both sides, as illustrated in Fig. 148.

Gullies

Where there is a combined sewer that takes both soil and waste water discharges and also rainwater from roofs and paved areas, it is necessary to use a trapped gully at the foot of rainwater downpipes so

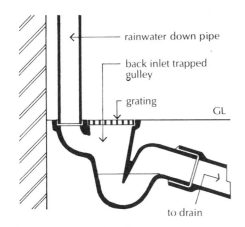

Fig. 149 Trapped gully.

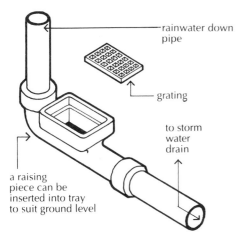

Fig. 150 Rainwater shoe.

Private sewers, common drain

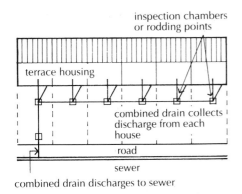

Fig. 151 Common drain.

that the water seal in the gully serves as a barrier to foul gases rising from the drains. The trapped gully illustrated in Fig. 149 has a back inlet connection for the rainwater pipe and a grating that serves as access to clear blockages and to take water running off paved areas. These gullies are made with either back or side inlet connections for rainwater pipes.

An alternative arrangement is to use a trapped gully without either back or side inlet and to fix the rainwater pipe so that it discharges, through a shoe, over the gully grating. The disadvantage of this is that the rainwater discharge may splash water around the gully and cause damp patches on walls. To drain paved areas to a drain discharging to a combined sewer, a trapped gully is used.

Where there are separate drain and sewer systems for foul water and rain and surface water, the gullies that collect rainwater and surface water can be connected directly to the surface or storm water drain, without trapped gullies, as there are no foul gases or air in the drain or sewer that may cause a smell.

The fitting used to collect rainwater (Fig. 150) is described as a rainwater shoe to differentiate it from gullies that have a water seal. The shoe has a grating that fits loosely into a tray to serve as access to clear blockages and collect water from paved areas.

As there is less likelihood of blockages in surface water drains than in soil water drains, it is not considered necessary to form inspection chambers at all junctions, bends and changes of gradient to the drain as is the case with foul-water drains. Rodding eyes at salient points to facilitate clearing drains are generally considered adequate for the purpose.

Because of the wide range of fittings available it is common to use clayware gullies and rainwater shoes for most drain pipe materials. Gullies and shoes are usually bedded on a small concrete base to provide a solid base for connection to stack pipes and drains.

Considerable economies in drainage costs may be effected by the use of combined drains or private sewers. There is no exact distinction between the words drain and sewer, but the most generally accepted definition is that pipelines under privately owned land, laid and maintained by the owner, are called drains, and pipelines laid and maintained by the local authority under roads are called sewers.

A private sewer, also termed a combined drain, is a system of drains laid for the use of two or more buildings, paid for and maintained by the owners of the buildings and making one connection to the public sewers. The reasons for laying a private sewer or combined drain are economy and convenience.

A common drain or sewer to a terrace of houses such as that illustrated in Fig. 151 requires one drain connection to the public sewer.

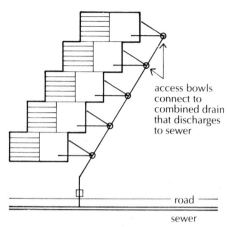

access bowls
connect to
combined drain
that discharges
to sewer

road

sewer

Fig. 152 Combined drains (private sewers).

The local authority makes a charge for a connection to its sewer, which not inconsiderable cost may be shared by several householders. This initial cost saving should be set against the cost and inconvenience of disputes over who should pay for subsequent repairs and maintenance.

Where properties do not directly front on to public roads and sewers, as is common in houses such as those shown in Fig. 152, it is plainly convenient to lay a private sewer by use of a combined drain, making one connection to the public sewer rather than separate connections from each house. Here economy and good sense combine. Having laid a private sewer or combined drain it is necessary to apportion the cost of maintenance work to it. Here lies the source of inevitable dispute over who should pay what.

If one householder were to damage the length of private sewer that ran under his land, should all pay a portion of the cost of repair or just the one owner? How to prove that the one owner caused the fault, and the 'falling out' of neighbours, is a possible consequence of the system.

Connections to sewers

Connections to public sewers are generally made by the local authority and paid for by the building owner or made by the building owner under the supervision of the local authority. In new developments with new sewers a branch connection is constructed in the sewer and where there is an existing sewer the old sewer is broken into and a new connection is made.

Intercepting traps

Until comparatively recent times, it has been practice, in the older urban areas, to build an intercepting or disconnecting trap into the drainline connection to combined foul and rainwater water sewers. An intercepting trap is a water-sealed trap incorporating a rodding eye which is built into an intercepting chamber near the boundary of buildings in the outfall drain to sewers, as illustrated in Fig. 153. The purpose of the trap is to provide a water seal between the sewer and the private drain against sewer gases rising into the private drains, and hopefully also as a bar against rats finding their way from the sewer into the private drain system.

Fresh air inlet (FAI)

Because of this water seal, a ventilation pipe, termed a fresh air inlet (FAI), is connected to the inspection chamber to ventilate the private drain, as illustrated in Fig. 153. An intercepting trap is a prime cause of blockages in foul drains. It is no longer considered necessary to provide a seal between sewer and drains, and modern practice is to

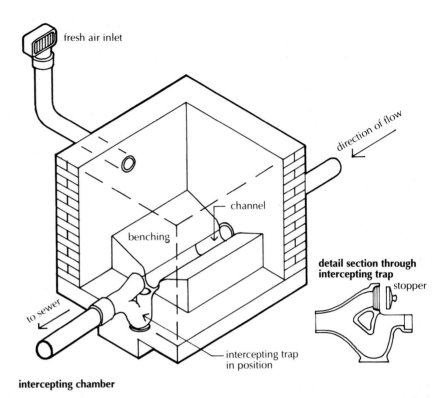

Fig. 153 Intercepting manhole.

intercepting chamber

dispense with the intercepting trap and discharge the drain directly into the sewer.

DRAIN TESTING

Water test

All newly laid drainlines should be tested for watertightness after jointing and laying and again after backfilling and consolidation of trenches. The drain is tested by water pressure applied by stopping the low end of the drain and filling it with water to a minimum head of water, as illustrated in Fig. 154. The head of water should be 1.5 m above the crown of the high end of the pipeline under test. Where long drainlines are to be tested, and the head of water would exceed 6 m due to the length and gradient of the drain, it is necessary to test in two or more sections along the line.

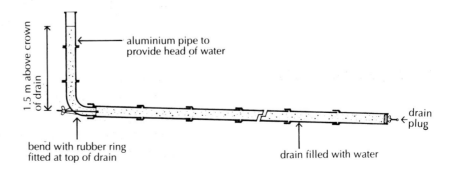

Fig. 154 Water test.

The loss of water over a period of 30 min should be measured by adding water from a measuring vessel at regular intervals of ten minutes, and noting the quantity required to maintain the original water level. The average quantity added should not exceed 1 litre per hour per linear metre, per metre of nominal internal diameter. The water test is a test of watertightness under pressure, a condition that a freely flowing drain will never suffer, and is thus a test of water-tightness far more rigorous than the drainline is designed for.

Air test

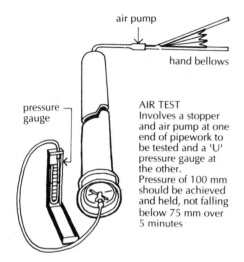

air pump

hand bellows

pressure gauge

AIR TEST
Involves a stopper and air pump at one end of pipework to be tested and a 'U' pressure gauge at the other.
Pressure of 100 mm should be achieved and held, not falling below 75 mm over 5 minutes

Fig. 155 Air test.

The air test is generally accepted as a less rigorous test than the water test. The drainline to be tested is stoppered at both ends and air pressure is provided by a pump, the pressure being measured by a graduated U tube or manometer, as shown in Fig. 155.

Expanding drain stoppers suited to take the tube from the pump at one end of the drainline and the tube to the pressure gauge or manometer at the other, are fitted to the ends of the drainline to be tested.

Air pressure is applied to the drain through the hand or foot operated pump until a pressure of 100 mm is recorded on the graduated U-section pressure gauge and the valve to the gauge is shut. An initial period of 5 min is allowed for temperature stabilisation and the pressure is then adjusted to 100 mm. The pressure recorded by the U gauge or manometer should then not fall below 75 mm over a further 5 min for a satisfactory test.

Where trapped gullies and ground floor appliances are connected to the drainline being tested, a pressure of 50 mm is adopted as the measure.

Smoke test

The smoke test has been used for old drains where the water or the air test is too rigorous for them. The drain to be tested is stoppered at suitable intervals and smoke is introduced under pressure from a smoke capsule or smoke machine. The purpose of the tests is to discover leaks by the escape of smoke, either when the line has been uncovered or is underground. An escape of smoke will find its way to the surface through a considerable depth of soil and all but the most dense concrete cover. Fig. 156 illustrates a smoke test machine.

The reason for using a smoke test, particularly on old drains, is to

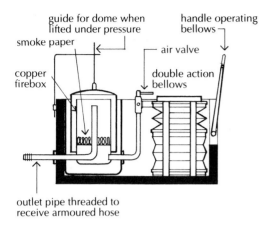

guide for dome when
lifted under pressure

handle operating
bellows

smoke paper

air valve

copper
firebox

double action
bellows

outlet pipe threaded to
receive armoured hose

Fig. 156 Eclipse smoke testing machine.

CCT surveys

SEWAGE COLLECTION AND TREATMENT

Cesspools

give some indication of the whereabouts of likely leaks before any excavation to expose drains has been undertaken. This may be of use on long runs of drain and where drains run under buildings. The appearance and smell of smoke may give a useful indication as to where excavation to expose drains should take place.

Water, air and smoke tests act on the whole of the internal surface of drains. These tests may indicate leaks due to cracks in the crown of the drain. As drains never run full bore a crack in the crown of a drain may not cause any significant leakage and may not warrant repair.

Modern technology provides a means of making visual inspection of the inside of drain runs by the use of a small camera that is inserted into and run down the line of a drain to provide a moving picture record on a monitor of the view of the inside of the drain. This is usually recorded on a cassette or disc. This somewhat expensive survey may well be a worthwhile alternative to excavation to expose drains.

With the introduction of sanitary water closets (WCs), a cesspool was the traditional means of collecting the discharge from WCs. At the time and up to comparatively recently, it was considered that only the liquid outflow from WCs was insanitary. Water from washing and cooking was not deemed to be insanitary and was treated as dirty water.

In consequence waste water from baths, basins and sinks was drained to soakaways along with rainwater from roofs and paved areas, and later in the then developing urban areas, to combined sewers that collected all liquid discharges alike. At the time society was much less regulated than it is now and less fearful of infections from the natural cycle of life and decay.

A cesspool is an underground chamber or container used to collect and store all foul and waste water from buildings. They are emptied by pumping the contents to a tanker.

Today cesspools are only used in outlying areas where there is no ready access to a sewer and where ground is waterlogged or the slope of the site does not allow the use of a more compact and convenient septic tank.

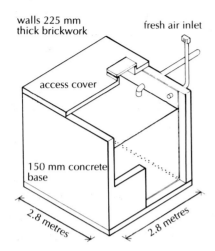

walls 225 mm
thick brickwork

fresh air inlet

access cover

150 mm concrete
base

2.8 metres

2.8 metres

Fig. 157 Cesspool.

The size of the cesspool depends on an assumption of the amount of water used each day per person at 137 litres per head, and the number of days' storage capacity required before the cesspool can be drained. A minimum capacity of 18 000 litres capacity is recommended for an average household where the cesspool is drained every 45 days.

Many local authorities provide a service of emptying cesspools on a regular basis, weekly or monthly, free of charge, and more frequent collections on payment by the building owner. The authorities empty the cesspools by pumping to a tanker which then discharges the foul liquid to a sewage treatment works.

To comply with current regulations a cesspool should have no outlet and should effectively be watertight to loss of water from within and entry of ground water from without.

The traditional cesspool was a brick lined pit, such as that illustrated in Fig. 157, with engineering brick walls 220 mm thick or well burnt dense bricks lined inside with cement and sand rendering on a concrete base. A reinforced concrete cover over the cesspool would be formed at, or just above, ground level with some form of access cover for emptying and a fresh air inlet.

A range of prefabricated cesspools is available today. These are made from glass reinforced fibre plastic (GRP) in the form of sealed containers with capacities of 7500 to 54 000 litres. Larger capacities of up to 240 000 litres can be made.

The ribbed cylindrical cesspool, with access point, inlet and fresh air inlet (FAI), is delivered ready to lower into an excavated pit with the cylinder lying on its long axis. The cesspool is laid on a bed of concrete and surrounded with a lean mix of concrete, with the cover to the access point at or just above ground level.

Septic tank

A septic tank differs from a cesspool in that a cesspool is designed to contain and retain all the outflow of soil and water, whereas a septic tank is designed to take the outflow of soil and waste water, retain some solid organic matter for partial purification, and discharge the liquid sewage through a system of land drains to the surrounding ground to complete the process of purification.

A septic tank is an underground container whose function is to collect all the soil and waste water discharge, and inside the tank a series of baffles slow the flow of liquid waste from the inlet to the outlet and cause settlement of larger sewage particles to the base of the tank. The liquid content that is now partially purified then runs to a system of perforated or porous land drains from which it soaks into the ground for further purification. The extent of these so-called leaching drains depends on the nature of the soil, its porosity, the water table and the capacity of the septic tank.

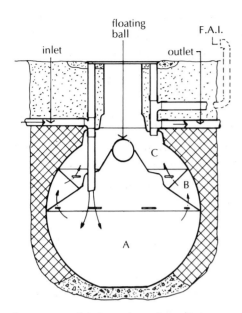

Fig. 158 Prefabricated septic tank.

The usual capacity of septic tanks is from 2800 to 6000 litres for from 4 to 22 people respectively. A range of moulded, high grade polyethalene, bulb-shaped tanks is manufactured, with access turret, inflow and outflow connections and a fresh air inlet (Fig. 158).

The tank is delivered ready to lower into an excavated pit. The tank is bedded on concrete and surrounded with lean mix concrete with the access cover at or just above ground level.

The septic tank is divided into three chambers. The outflow of the foul and waste water drains is discharged to the lower chamber (A in Fig. 158) in which the larger solid particles of sewage settle to the bottom of the chamber. As more sewage flows down the inlet, liquid rises through the slots in the bell shaped division over the lower chamber and enters the middle chamber (B). In this chamber fine particles of solid matter settle on top of the bell shaped division and settle through the slots into the lower chamber.

As the tank fills, liquid now comparatively free of solid sewage matter rises through the slots in the inverted bell shaped division to the third and upper chamber (C). As this upper chamber fills, the now partially purified liquid rises through the outflow to a system of perforated land drains.

The land or leaching drains, laid in trenches surrounded by granular material, spread the liquid sewage over an area sufficiently large to encourage further purification of the liquid sewage by aeration and the action of micro-organisms in sewage.

The accumulated sludge of solid sewage that settles in the base of the tank should be removed at intervals of not more than 12 months. This is effected by a tanker. A hose from the tanker first empties the upper chamber, by suction. As the liquid level falls, the ball that seals the division between the upper and middle chambers falls away. This provides access to remove the sludge from the lower chamber.

Sewage treatment plant

The preferred and more expensive system for sewage outflow, where no sewer is readily available, is the installation of a sewage treatment plant that will produce a purified liquid outflow that can be discharged to nearby ditches and streams without causing pollution of water supplies. A form of sedimentation tank, similar to a septic tank, causes solid sewage to settle. The resultant liquid outflow is further purified by exposure to air to accelerate the natural effect of micro-organisms, native to sewage, combining with oxygen to purify the sewage.

The three systems used to speed the exposure of the liquid sewage to air are:

(1)   By spreading over a filter bed
(2)   By spreading over rotating discs
(3)   By pumping in air to combine with the liquid (aeration)

The traditional sewage treatment plant, such as that illustrated in Fig. 159, comprises a settlement tank which acts in the same way as a septic tank to allow solid matter to sink to the bed of the chamber either naturally or assisted by baffles. This first treatment generates a thin film of scum (biomass) on the surface of the liquid, which acts as a seal against air to encourage anaerobic micro-organisms to break down the solid particles.

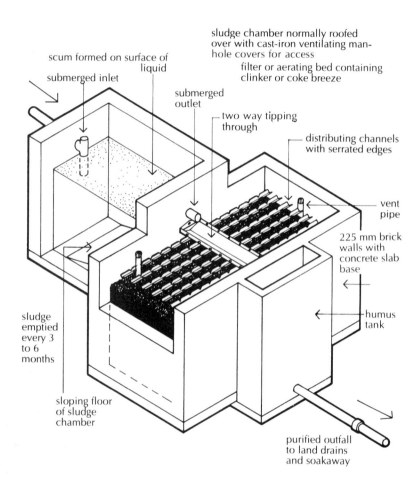

Fig. 159 Sewage treatment plant.

Liquid then flows from the settlement tank to a two-way tipping trough into distributing channels which spread the liquid over and through the filter medium. The filter medium is clinker or coke breeze chosen to expose the liquid to the maximum surface area of air. The then purified liquid runs through a humus tank to drains or a soakaway.

The small treatment plant shown in Fig. 159 is similar in operation to the larger filter beds used in municipal sewage works. It is built on a concrete base in an excavated pit, with brick walls lined inside with cement and sand rendering. The plant is roofed with reinforced concrete, with cast iron manhole covers inset for access for periodic cleaning.

The advantage of the small traditional sewage treatment plant is that there are no mechanically operated moving parts requiring attention. The disadvantages are that the plant is somewhat laborious to construct and the filter bed will need cleaning to remove sludge if it is to function efficiently.

Prefabricated sewage treatment plant

A variety of prefabricated sewage treatment units is available to service single and multiple groups of houses. These units are made from high grade moulded polyethalene. The units are delivered to site ready to be lowered into an excavated pit where they bear on a concrete bed and are surrounded with lean mix concrete. The standard units which are made for the treatment of sewage from 5 to 300 people have primary tank capacities of from 1100 to 40 000 litres.

The two systems used in these prefabricated units are the introduction of highly aerated water which is pumped into the treatment chamber by an electrically operated pump, or the use of several slowly rotating discs that expose the liquid sewage to air. The aim in both is to expose the liquid sewage to the maximum contact with air to accelerate treatment for purification.

Like the purpose-built sewage treatment plant in Fig. 159, the prefabricated treatment plant in Fig. 160 consists of three stages: a

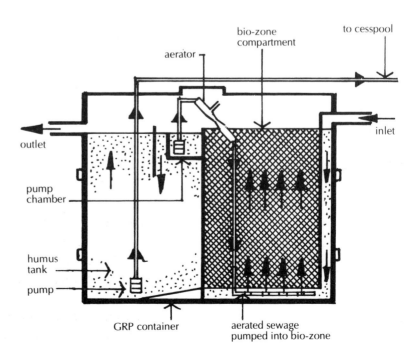

Fig. 160 Prefabricated sewage treatment plant.

settlement tank, a filter or Bio-zone and a humus tank. The settlement tank may be an existing brick-built septic tank or one of the bulb shaped GRP settlement tanks such as that in Fig. 158.

The filter or Bio-zone and humus tank are contained inside a ribbed GRP (glassfibre reinforced plastic) unit. In the filter or Bio-zone compartment, closely packed plastic coils or a honeycomb are supported and separated from the ends and base of the GRP unit. Liquid sewage from the settlement tank flows down a vertical channel to the base of the filter medium, through which it rises by natural gravity up to the liquid level of the filter and out over a weir to the pump chamber and humus tank. An electrically operated pump, submerged in the liquid sewage in the small pump chamber, forces liquid up a pipe, through an aerator and down to a perforated pipe below the filter. In this way the aerated liquid sewage rises through the filter to encourage and accelerate the action of aerobic micro-organisms in purifying the sewage.

A baffle in the humus tank encourages liquid to flow down to the bed of the tank and then up to the outlet. In so doing, bacterial cells group together to form larger particles, known as flocks, to sink more quickly to the base of the humus tank. An electrically operated recycling pump in the base of the humus tank returns the settled flocks back to the inlet of the settlement tank to cause further purification to take place.

The two electrically operated pumps in the pump chamber and the humus tank operate continuously to return the liquid sewage to the filter or Bio-zone and the settlement tank for further treatment. The now purified liquid flows from the humus tank to a drain which discharges it to a watercourse, as it meets the high standards of purity required by the National Rivers Authority.

The treatment plant is lowered into an excavated pit in the ground and bedded on and surrounded with concrete so that the cover of the plant is at, or just above, ground level. Metal access covers are set in place and an electricity supply run to the pumps from a control unit fixed in some convenient place, such as a garage. The inflow and outflow drains are connected.

The sludge that collects in the base of the settlement tank is pumped out two or three times a year. Regular maintenance of the pumps is necessary.

Pollution control

Before a cesspool, septic tank or sewage treatment plant is installed it is necessary to make sure that the installation conforms to the requirements of the National Rivers Authority (NRA) in England and Wales and the River Purification Boards (RPBs) in Scotland. The requirements of these and other water authorities are usually overseen by local authority environmental control officers who will advise on requirements best suited to the locality they oversee.

Pumps for sewage

Where sanitary fittings and their drains are below the level of the sewer, due to the slope of the ground, or in basements, it is necessary to raise the foul water by pumping. A sewage pump is an expensive piece of equipment and requires frequent maintenance if it is to function adequately. The need for pumping sewage should thus be avoided if at all possible.

The types of equipment used are the pneumatic ejector and the mechanical pump. The pneumatic ejector is used for small installations where the flow of sewage is small, as for one or a few houses, and where the sewage has to be raised comparatively small distances.

The pneumatic ejector

The pneumatic ejector is a relatively simple device with few moving parts to go wrong. The sewage enters the ejector cylinder through a non-return valve and raises a float that operates the air valves. Compressed air forces the sewage out through another non-return valve up to the sewer level and as the float falls the compressed air is evacuated. The ejector cylinder is fed by gravity from the drains. The air compressor and its air cylinder may be fitted adjacent to the ejector unit or at a higher level. The air compressor automatically operates to keep the air cylinder charged.

The operation of the pneumatic ejector is simple and straightforward and requires the least maintenance.

Where there is appreciable flow of sewage to be raised some distance to the sewer, a mechanical pump system is used. The two systems used are the submersible pump and the dry well suction pump.

Submersible pump

The submersible pump and motor unit is submerged in the foul sewage to be raised, or the pump is submerged with the motor raised, as illustrated diagramatically in Fig. 161. In either case maintenance of the submerged parts is a disagreeable task.

The motor and pump assembly is suspended in a concrete sump into which the liquid sewage from the drains flows. The electrically operated pump is activated by a float switch through a ball that floats on the liquid sewage. As the liquid rises, the float operates the switch to start the motor and pump, which operate until the liquid level falls. The liquid sewage is pumped up the outflow, and through a non-return valve to sewer level.

This pumped sewage system requires fairly frequent maintenance and cleaning to ensure that the motor, pump, switch and non-return valve operate satisfactorily. The advantage of the submersible pump is that it operates more efficiently than a pump distant from the liquid to be pumped.

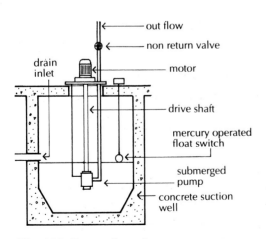

Fig. 161 Dry well suction pump.

Dry well suction pump

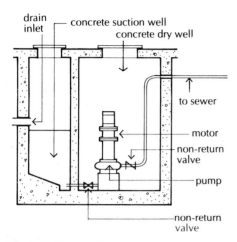

Fig. 162 Dry well sewage pump.

The dry well sewage pump is installed in a dry chamber or well adjacent to the chamber containing the sewage to be raised. An electrical motor and pump assembly is fixed in an underground chamber which is adjacent to a well into which the sewage outgo flows from the drains. The liquid sewage is pumped from the well, through non-return valves, up to the level of the sewer. Some form of float switch in the well controls the operation of the motor and pump to limit pumping operations.

The dry well is sufficiently large for an operative to climb down into the dry well for maintenance and repairs. Fig. 162 shows a typical layout. For continuous operation it is necessary to duplicate sewage pumping systems.

5: Roof and Surface Water Drainage

ROOF DRAINAGE

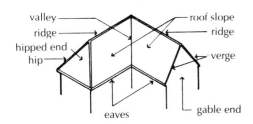

Fig. 163 Roof drainage.

Rainwater running off both pitched and flat roofs is usually collected by gutters and outlets and discharged by rainwater downpipes to drains, sewers or soakaways. The reason for collecting rainwater from pitched roofs is that unless the eaves have an appreciable overhang, the rainwater might saturate walls, particularly when the falling water is wind blown against wall surfaces.

Rainwater gutters and downpipes are a prime source of dampness in walls. Gutters that are blocked, cracked or have sunk may cause persistent saturation of isolated areas of wall, leading to damp staining and possible dry rot in timbers. Gutters are not readily visible to building owners and are generally difficult to access for regular clearing of leaves and other accumulated debris.

The terms used to describe the various parts of pitched roofs are given in Fig. 163. 'Eaves' is used to describe the general area of the lower edge of a pitched roof that drains to an eaves gutter. 'Verge' is that edge of a roof slope that finishes or verges over a gable end wall. Here the roof covering is raised somewhat to discourage wind blown rainwater from running down the gable end.

A valley is formed by the junction of two roof slopes at right angles where the rainwater running down the two slopes meets to run down to eaves. Valleys are particularly liable to blockages. They are difficult to inspect and difficult to access and are generally disregarded until a leak occurs.

The size of gutters and down pipes is determined by the estimated volume of rainwater that will fall directly on a roof during periods of intense rainfall. It is practice to use gutters and rainwater downpipes large enough to collect, contain and discharge water that falls during short periods of intense rainfall that occur during storms. Rainfall intensities of 75 mm/hour occur for 5 min once in every 4 years and for 20 min once in every 10 years. It is for such periods of intense rainfall that the drain system is designed.

To determine the amount of water that may fall on a pitched roof it is necessary to make allowance for the pitch or slope of the roof and for wind driven rain that may fall obliquely on a slope of roof facing the wind direction. A larger allowance for wind driven rain is made for rain falling on steeply sloping roofs than on roofs with a shallow slope. The effective area of roof to be drained is derived from the plan (horizontal) area of a roof multiplied by a factor allowing for the pitch or slope of the roof, as set out in Table 12.

Table 12 Calculation of area drained.

Type of surface	Design area [m²]
flat roof	plan area of relevant portion
pitched roof at 30°	plan area of portion × 1.15
pitched roof at 45°	plan area of portion × 1.40
pitched roof at 60°	plan area of portion × 2.00
pitched roof over 70° or any wall	elevational area × 0.5

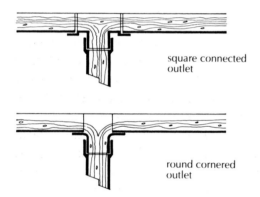

square connected outlet

round cornered outlet

Fig. 164 Rainwater outlets.

Having determined the effective area of a roof slope to be drained it is necessary to select a size of gutter capable of collecting and discharging the volume of water assumed to fall during storms. For the purpose of choosing a gutter size it is assumed that the roof drains to a half round gutter up to 8 m long with a sharp edged outlet at one end only and laid level. This supposes an extreme condition as a run of eaves gutter 8 m long is unusual. This extreme condition is adopted on the grounds that if it can cope with rainwater collection all other less extreme conditions will operate successfully.

Most gutters are laid with a slight fall to encourage flow, and the majority of plastic eaves gutters have round cornered outlets to encourage discharge. Square cornered gutter outlets, such as those in cast iron, impede flow, whereas plastic gutters, which commonly have round edged outlets, cause less impedance to flow, as illustrated in Fig. 164. In the calculation of rainwater downpipe sizes some reduction of pipe size may be effected by the use of round edged gutter outlets.

To make the best use of the gutter it is practice to fix it to fall towards each side of an outlet to economise in gutter size and number of rainwater pipes, as illustrated in Fig. 165. Obviously to drain a given roof, a large gutter will require fewer outlets and pipes than a small gutter, as illustrated in Fig. 165. Which of the two arrangements shown is used will depend on economy in use of gutters and downpipes, economy of drain runs, position of windows and appearance. In the examples shown in Fig. 165 it would be possible to utilise two rainwater pipes, one at each end of the roof, with the gutter falling each way from the centre of the roof. Each half of the length of gutter would have to collect more rainwater than any one length of fall shown in Fig. 165, and a large gutter would be required.

The length of a gutter between outlets to rainwater downpipes will depend on the position and number of rainwater pipes related to economy and convenience in drain runs, which in turn will determine the area of roof that drains to a particular gutter length or lengths.

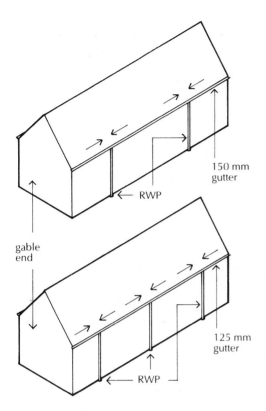

gable
end

Fig. 165 Gutters and downpipes.

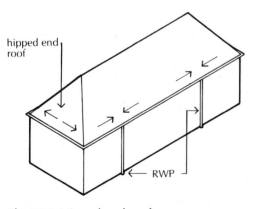

Fig. 166 Hipped end roof.

Table 13 gives an indication of the required diameter of half round gutter sizes related to the maximum effective area of roof that drains to that gutter, where the gutter is fixed level and the outlets are square edged.

Table 13 Gutter sizes and outlet sizes.

Max roof area [m^2]	Gutter size [mm dia]	Outlet size [mm dia]	Flow capacity [litres/sec]
6.0	—	—	—
18.0	75	50	0.38
37.0	100	63	0.78
53.0	115	63	1.11
65.0	125	75	1.37
103.0	150	89	2.16

Refers to half round eaves gutters laid level with outlet at one end sharp edged. Round edged outlets allow smaller downpipe sizes.

The figure given for flow capacity in gutters is given in litres per second, which is derived from rain falling on the relevant effective area of roof in millimetres per second where a cubic metre of water equals 1000 litres. This figure may be used to check gutter sizes chosen against manufacturers' recommended flow capacities of gutters.

The gutter outlet sizes will determine the size of rainwater downpipe to be used. For the majority of small buildings, such as single houses and small bungalows, a gutter with a diameter of 75 or 100 mm is adequate for the small flows between outlets.

Eaves gutters to hipped roofs are fixed to collect rainwater from all four slopes with angle fittings at corners, as shown in Fig. 166, to allow water to run from the hipped end slope to the outlets to main slopes. In Fig. 166 two downpipes are shown to each main roof slope, with the gutter to the hipped ends draining each way to the gutters to the main roof slopes. A square angle in a gutter, within 4 m of an outlet, somewhat impedes flow, and allowance is made for this in the calculation of flow in gutters.

Flooding and overflow of eaves gutters to hip ended roofs most commonly occur at angles where obstruction to the flow is greatest, blockages are most likely to occur and the angle gutter fittings may sink out of alignment.

The position of rainwater pipes will depend principally on economy and convenience in making connections to drains, rather than an ideal arrangement of gutters and their outlets. Running lengthy and complicated drain runs is more expensive that adjusting gutter outlets and rainwater drain pipes to suit economical drain runs.

segmental gutter

half round gutter

ogee or OG gutter

Fig. 167 Rainwater gutter cross-sections.

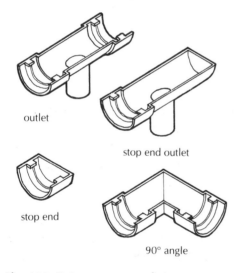

outlet

stop end outlet

stop end

90° angle

Fig. 168 Rainwater gutter fittings.

Eaves gutters to most roofs are fixed to a shallow fall of 1:360 towards outlets. This shallow fall avoids too large a gap between the edge of the roof covering and the low point of the gutter, yet is sufficient to encourage flow in the gutter and make allowance for any slight bow or settlement of the gutter.

The section of gutter most commonly used for the traditional pitched roof covered with slates or tiles is half round. This is the standard section of the generally used plastic and the traditional cast iron gutter. Other sections are available, such as the segmental and ogee or OG gutter illustrated in Fig. 167.

The segmental may be used to collect rainwater from small shallow pitched roofs. The ogee section was much used for cast iron gutters for its appearance as a feature to larger, more imposing houses and other buildings.

Gutter lengths and their associated fittings have spigot and socket ends so that the plain (spigot) end of gutter fits to the shaped (socket) end of gutter to provide a level bed of gutter. The necessary fittings to gutters are running outlets, stop end outlets, angles both internal and external, and stop ends. These fittings have socket ends to fit the spigot ends of the section of gutter.

Fig. 168 shows typical plastic gutter fittings for half round gutters. The angle fitting shown is an external angle fitting. Internal angle fittings are used where a wing or part of a building butts to another part at 90°.

Standard half round, uPVC gutter sections with spigot and socket ends and angles, outlet and stop end fittings, are commonly used to drain the majority of pitched roofs today. uPVC is a lightweight material that needs no protective coating, is moderately rigid and has a smooth surface that encourages flow. It is usually made with a black or pale grey finish. This material is particularly used for its comparatively low cost and freedom from maintenance.

Fig. 169 shows uPVC gutters. Flexible seals are bedded in the socket ends of both gutters and fittings. These seals are watertight and at the same time are sufficiently flexible to accommodate the appreciable thermal expansion and contraction that is characteristic of this material. Without the flexibility of the seals, long lengths of gutter might otherwise deform. The joint between spigot and socket ends is secured with a gutter strap clipped around the gutter.

uPVC gutter lengths are supported by plastic fascia brackets that are screwed to fascia boards. The gutter is pressed into the fascia brackets so that the lips of the bracket clip over the edges of the gutter to keep it in place. The spacing of the brackets is determined by the section of gutter used.

Rainwater downpipes are moulded with plain spigot and shaped socket ends. Socket ends are shaped to make the close fitting joint to

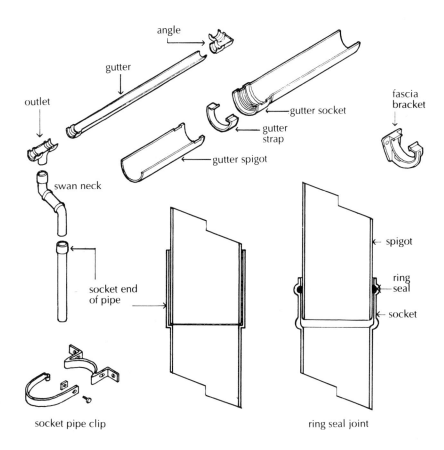

Fig. 169 uPVC rainwater gutters and downpipes.

spigot ends, or moulded to take a flexible ring seal as illustrated in Fig. 169. The former makes a reasonably close fit and the latter a more positive watertight fit. The shaped fitting or fittings from the gutter outlet to the downpipe are formed to make allowance for overhanging eaves so that water discharge is directed from the gutter towards the downpipe fixed to walls.

The swan neck fittings are either moulded in one piece to accommodate particular dimensions of eaves overhang, or consist of three fittings to allow for various eaves overhangs. The swan neck shown in Fig. 169 comprises three units: two bends and a short straight length used to allow for a particular eaves overhang.

Rainwater downpipes are secured to walls with plastic pipe clips screwed to walls. The two piece pipe clip illustrated in Fig. 169 facilitates fixing plugs and screws to walls.

Cast iron rainwater gutters and downpipes

Cast iron was the traditional material for gutters and downpipes for pitched roofs. It is heavy, rigid, brittle and durable as long as it is protected by paint. Standard gutters are of half-round or ogee section with socket and spigot ends with angle, outlet and stop end fittings.

The advantage of cast iron as a gutter material is that it is much less likely to sag than uPVC. Its disadvantage is that it is liable to rust and should be coated to inhibit rusting and for appearance sake.

The gutter joints are bedded in red lead putty or mastic and bolted together. Gutter lengths are supported by fascia or rafter brackets for half-round gutters and by screws or brackets for ogee gutters. Fig. 170 shows a cast iron gutter and downpipe.

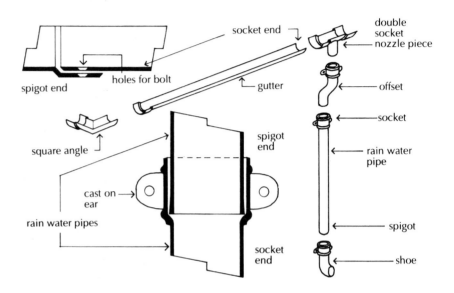

Fig. 170 Cast iron rain water gutter and pipe.

Standard round-section socket and spigot downpipes with cast-on ears for fixing are used with offset and shoe fittings. The downpipes are secured to walls with pipe nails driven into hardwood plugs through distance pieces to fix the pipes away from the wall for painting.

There have been two schools of thought as to whether the joints between downpipes should be left open or sealed. One claimed that open joints would give an indication of blockages in pipes by water overflowing from a particular joint, and so would indicate which pipe length to unblock. The second claimed that all joints should be sealed to prevent any seepage of water from them. Open joints are probably more useful with cast iron pipes whose coarse textured surface is more conducive to blockages than the smooth surface of uPVC pipes.

Aluminium, galvanised and enamelled finish pressed steel and zinc have been used for gutters and downpipes. None of these materials are now much used.

Gutters to wide span roofs

The considerable volume of rainwater that will fall on pitched roofs over extensive buildings such as factories, warehouses and sheds has to drain to gutters on one or more boundaries of the building, or to valley gutters between pitched roofs. Because of the considerable volume of rainwater that may fall on such roofs during periods of intense rainfall it is generally necessary to make an estimation of likely rainfall and gutter sizes needed to collect and discharge the water.

The estimation of water volume and gutter and downpipe sizes, described earlier in this chapter, is plainly more relevant to large roofed areas than to a small house where the standard 100 mm gutter will be adequate.

The section of gutter used is described as a boundary wall or box gutter because the gutter is bedded in or fixed to a boundary wall and is usually box-shaped. The material used for these gutters is pressed steel either galvanised or more usually galvanised and plastic coated to inhibit rust. Some typical sections are shown in Fig. 171. The socket and spigot ends of gutter are bedded in mastic and bolted together. Various outlet, stop end and angle fittings are made.

These gutters are either bedded on a boundary wall, supported by metal brackets as a verge gutter or used as a parapet gutter. The gutters are laid or fixed level with as few outlets as possible to discharge the estimated flow.

Valley gutter

Adjacent pitched roofs drain to a common valley gutter which either discharges to a rainwater outlet or a boundary gutter where there are hipped ends to roofs. These gutters are either galvanised pressed steel or plastic coated galvanised pressed steel. Some sections are shown in Fig. 171. A wide bed of gutter is convenient for access to clear blockages. The effective area of roof to be drained is determined as in Table 12.

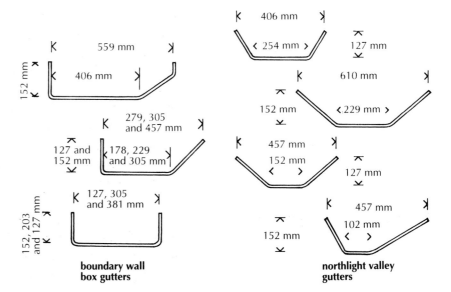

Fig. 171 Gutters.

boundary wall
box gutters

northlight valley
gutters

Flat roofs

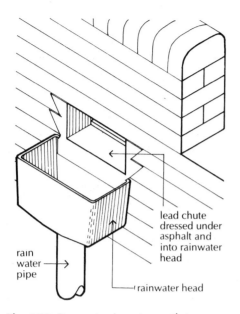

Fig. 172 Parapet rainwater outlet.

Connection to drains

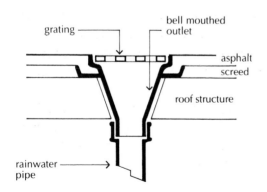

Fig. 173 Rainwater outlet.

Flat roofs should be laid to a shallow fall towards outlets to avoid what is termed 'ponding'. Ponding is caused by too shallow a fall or the inevitable deflection under load of any horizontal structure that will cause water to lie on a flat roof in the form of a shallow pond. The static ponded water will accelerate deterioration of most flat roof coverings and will penetrate cracks caused by thermal and moisture movements of the structure.

For appearance sake flat roofs are often surrounded by parapet walls raised above the level of the roof finish. The run off of rainwater is encouraged by a fall or slope of the roof to outlets formed in the parapet wall. These outlets are lined with lead sheet shaped to fit in the bed and side of the outlet and are dressed down over a rainwater head, illustrated in Fig. 172. The lead chute is dressed under the flat roof covering. Because the run off of rainwater is comparatively slow, fairly close spacing of outlets is necessary for drainage.

Rainwater heads are usually of galvanised or galvanised plastic coated pressed steel with an outlet to the rainwater downpipe. The traditional cast iron rainwater head is little used because of its cost and the need for regular coating or painting to avoid unsightly rusting. Where a flat roof is carried over boundary walls it discharges rainwater to an eaves or boundary wall gutter.

Large expanses of flat roof should drain to outlets in the roof as well as boundary outlets to avoid extensive lengths of fall or slope to flat roofs. Outlets are formed in the roof in some position where the necessary rainwater downpipe may be fixed to part of the supporting structure, such as a column. These outlets are shaped so that the roof covering may finish to the outlet with a watertight joint. Fig. 173 is an illustration of an outlet to a flat roof covered with asphalt, with a lift-up grating to facilitate clearing blockages.

It used to be practice to discharge rainwater from downpipes through a rainwater pipe shoe over a gully. This is not considered good practice today because water splashing from the shoe is liable to make the adjacent wall damp. Practice today is to connect down pipes to a back inlet gully, trapped gullies being used where the drainage is a combined drain system, and untrapped gullies where separate drain systems are used. Fig. 149 shows the connections of downpipes to gullies.

SURFACE WATER DRAINAGE

Paved areas

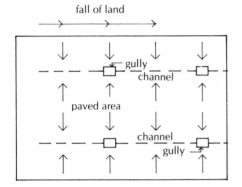

Fig. 174 Paved area drainage.

Channels

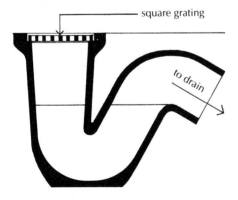

Fig. 175 Clay yard gully.

External surfaces that are paved with concrete, natural or artificial paving slabs, bricks or granite sets, should be laid with slight slopes or falls to gullies or channels. The purpose of the falls or slopes to drains is to discharge rainwater reasonably quickly for the convenience of people and to prevent ponding of water that would accelerate deterioration of paving materials by saturation and the effect of frost on water lying in fissures in the material.

The slope or fall of paved areas should be sufficient to drain water to outlets, yet not so steep as to make the surface slippery in wet or frosty weather. A minimum fall of 1:60 is generally recommended for paved areas on flat ground.

A fall or gradient of 1:30 to 1:20 is recommended along the length of access roads, with a crossfall of 1:40 across the width, usually formed as a camber or shallow curve to each side. As a general guide a paved area of 200 to 250 m^2 of paved area should drain to each gully at a slope, fall or gradient of 1:50. Paved areas abutting buildings should drain away from the building.

To economise in drain runs, paved areas should drain in two or more directions to yard gullies, with drainage channels between gullies in large paved areas to effect further economies. The drainage of the paved area in Fig. 174 is arranged by each way falls to channels that in turn drain to gullies. Where the fall or gradient is increased from 1:50 to 1:20 the effective area that can be drained to one gully can be increased to 500 m^2.

Channels between gullies may be level and depend on natural run off or may be laid to a slight fall by a gentle sinking towards a gully. A small square area of paving may be laid to fall to one central gully, with channels formed from each corner to the gully by the intersection of slopes. Forming this slope at intersections of falls, termed a current, may involve oblique cutting of paving slabs or bricks to provide a level surface.

Yard or surface water gullies are set on a concrete bed to finish just below the paved surface. The gullies generally have a 100 mm square inlet and are either trapped with a waterseal when they discharge to combined drains or are untrapped for discharge to a separate storm water drain.

Fig. 175 shows a trapped clay yard gully with loose cast iron grating for access to clear blockages. The graph in Fig. 176 indicates the area of paving that can be drained relative to estimated flow and size of drain relative to gradient or fall.

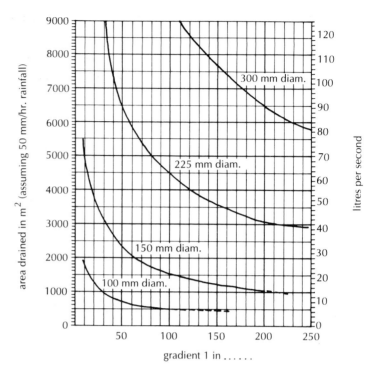

Fig. 176 Graph for determining diameter and gradient of surface water drains.

Drainage systems

Where there are separate drainage systems, roof and surface water drainage will connect to a separate system of underground drains discharging to the sewer. For the rain and surface water drains, it will generally be satisfactory to have a system of either clay or uPVC drains with rodding eye access fittings as necessary and an inspection chamber near the boundary. The description of pipe laying, bedding and gradient for foul water drains will apply equally to roof and surface water drains for which untrapped gullies are used, as there is no need for water seals and ventilation against foul odours.

Soakaways

Outside the more densely built-up urban areas there is often an adequate area of land as part of building sites to drain roof and surface water to soakaways.

A soakaway is a pit into which roof and surface water is drained and from which the water seeps into the surrounding ground. Soakaways are used where a combined sewer may not be capable of taking more water, where a separate rain and surface water sewer is distant and also as an alternative means of discharge.

Obviously if the soil is waterlogged and the water table or natural level of subsoil water is near the surface it is pointless to construct a soakaway. In this situation the surface water will have to be discharged by pipe to the nearest sewer, stream, river or pond.

In firm pervious ground such as chalk, it is generally sufficient to dig a pit into which the water is drained directly, and the pit is covered with a concrete slab as illustrated in Fig. 177.

In moderately compact soils such as clay, the pit may be filled with hardcore or clean broken stone to maintain the sides of the pit, as illustrated in Fig. 177.

In granular soils such as gravel and sand, the soakaway pit has to be lined with brick, stone or concrete to maintain the sides of the pit, and the lining of brick, stone or concrete must be porous or perforated to encourage the water to soak away into the surrounding soil, as illustrated in Fig. 177.

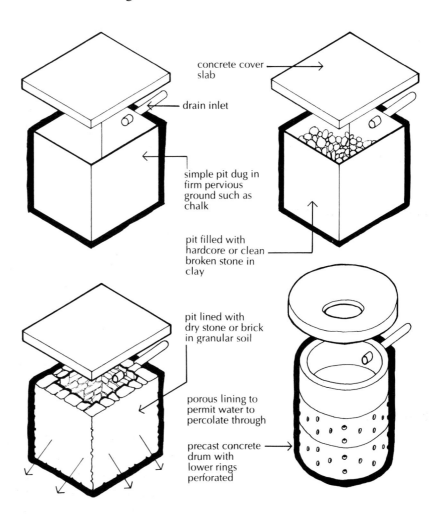

concrete cover slab

drain inlet

simple pit dug in firm pervious ground such as chalk

pit filled with hardcore or clean broken stone in clay

pit lined with dry stone or brick in granular soil

porous lining to permit water to percolate through

precast concrete drum with lower rings perforated

Fig. 177 Soakaways.

It may be cheaper to excavate two or more small soakaways rather than one large one, to reduce drain run lengths. Soakaways should be at least 3 m away from buildings so that the soakaway water does not affect the buildings' foundations, and they should also be on slopes down from buildings rather than towards buildings, to avoid overflowing and flooding.

6: Electrical Supply and Distribution

ELECTRICITY

The history of the development of a theory of electricity properly begins in the sixteenth century with William Gilbert, physician to Queen Elizabeth, who was first to use the word electric. He was studying the property of amber, after being rubbed, of attracting light objects, such as feathers. He named this power of attraction *electrica*, the Latin version of the Greek word for amber, a yellow, translucent, fossil resin much used in the manufacture of small ornaments.

Materials which have the property of attracting light objects, after being rubbed, were said to be electrified or charged with electricity. The energy used to overcome the frictional resistance between the two materials being rubbed together is converted to a charge of electrical energy in the amber. The more vigorously it is rubbed the greater the energy expended and the greater the charge of electricity in the amber.

Law of conservation of energy

Energy, which may neither be created nor destroyed (law of conservation of energy) is converted from one form – energy in rubbing to another, a charge of electrical energy in the amber.

Conductors, insulators

Early in the eighteenth century it was discovered that the electrical charge on amber could be transferred by contact with some materials and not with others. The materials through which the electrical charge could freely pass were called conductors and those through which the electrical charge could not pass were called insulators.

All metals are good conductors of electricity in that they offer small resistance to the transfer of electrical energy. Amber, sealing wax, hard rubber, dry glass and most plastics are good insulators in that they afford high resistance to the transfer of electrical energy.

The properties of copper and hard rubber or plastics as conductors and insulators respectively are used in electrical wire and cable. The plastic insulation around copper wire provides high resistance to the transfer of electrical energy from the conductive copper to another conductive material that touches the outside of the insulating plastic cover.

Earth, earthing

In the experiments to establish the properties of conductors and insulators, it was established that contact with an electrically charged material by hand had the effect of transferring the charge through the body to ground, as the human body is a fairly good conductor. The large mass of earth is said to be neutral electrically, so the human body conducts electrical charge by contact from the charged material to earth.

In the design of electrical installations, exposed metal, such as metal pipes, are connected to earth through a copper conductor as a path of least resistance to the discharge of electrical energy, rather than through the human body, as a safety device. This is described as the metal pipes being earthed.

Flow, current of electricity

In the middle of the eighteenth century it was suggested that electrical actions could be explained by supposing that electricity was a fluid. Electrically uncharged bodies contained a normal or equilibrium amount of the fluid and so showed no electrical effect.

When an electrically charged body was connected to a conductor some of the electric fluid flowed into the conductor, like a current of water flows in a stream. This analogy of the behaviour of water and electricity is expressed in the use of the word current, to define the sense of the transfer of electrical energy along a conductor. Some people still refer to 'turning on the juice', meaning turning on the electricity.

Amp

The unit of current, the ampere or amp, is the force of energy that one volt can move against one ohm of resistance in a conductor.

Resistance

All materials offer some resistance to the transmission or conduction of electricity by the conversion of electrical energy to heat. Some, like copper, have little resistance and others, such as plastics, have considerable resistance. In the design of electrical installations it is useful to know the measure of such resistance in order to choose the most suitable material for both conductors and insulators.

The unit measure of resistance to the transfer of electrical energy is the ohm, which is proportional to length and varies inversely to the cross sectional area of a conductor such as a copper wire.

For the most economical design of overhead cables for transmission of very high voltage electrical energy and the low voltage wiring in the home, the cross sectional area of a copper conductor relative to its resistance and cost per unit of length is a prime consideration.

Positive and negative electricity

Early in the eighteenth century two kinds of electricity were distinguished as vitreous and resinous electricity, the former generated by rubbing a glass rod and the latter by rubbing hard rubber. Bodies charged with vitreous electricity repelled one another and attracted those charged with resinous electricity, so establishing the rule that like repel and and unlike attract. Vitreous electricity is now called positive electricity and resinous electricity is called negative electricity.

In a body that is not electrically charged it is supposed that positive charges balance negative charges so that the body is said to be neutral electrically.

Voltage

Late in the eighteenth century Alessandro Volta (1745–1827) produced the first apparatus to provide a continuous current of electricity. His voltaic pile consisted of a pile of discs of copper and silver separated by discs of card or leather soaked in salt water. The reaction of the pile of copper, card, silver, card and so on, was to produce an electrical charge on the pile which caused a flow or current of electricity in a conductor connected to the two ends of the pile. This was the first battery capable of providing a stored charge of electricity that could be used to provide a continuous flow or current of electricity.

Volt

The charge of electrical energy stored in the pile may be given a numerical value in volts, which is the unit of potential electrical energy available. A volt is the unit of electrical energy or force necessary to cause one ampere of current to flow against one ohm of resistance.

Induced charges

Many experiments carried out in the eighteenth and early nineteenth centuries demonstrated that when an electrically charged conductor is brought near to, without touching, an uncharged insulated conductor, the insulated conductor acquires a charge that disappears when the charged conductor is removed. The action of causing a charge in a conductor by proximity is termed electrification by induction, and the charges are called induced charges.

The force between electric charges varies inversely to the square of the distance between them, so that the nearer the electrically charged wire is to the uncharged wire the greater the induced charge.

Magnetism

It had been known for many centuries that a particular mineral ore, lodestone, had the property of attracting iron and giving to the iron a like property of attracting other pieces of iron. The property of attracting is called magnetism, from the name of the district of Asia Minor where the lodestone was plentiful. This type of magnetism is described as ferro-magnetism.

Compass

During the twelfth and thirteenth centuries a property of iron, magnetised by being rubbed on lodestone, was first used as a compass. A thin rod or needle of ferro-magnetised iron, pivoted at its centre, will turn with one end towards the north pole of the earth no matter how often it is moved from that position. It is said that the magnetised needle is polarised, a term used in the later development of electro-magnetism.

Electro-magnetism

Up to the nineteenth century scientific enquiry into the nature of electricity, gravity, and ferro-magnetism continued largely as separate phenomena. Early in the nineteenth century, Oersted, a Danish scientist, demonstrated that an electric current flowing through a wire produced a magnetic effect on a compass needle by deflecting it to a position at right angles to the wire.

Michael Faraday, who had for some time been developing the theory that magnetism and electricity were of the same fundamental nature, began work on experiments to develop the idea. It was known that when an iron rod was moved in and out of a coil of electrically charged wire a charge was induced in the wire, and once the rod was still, the charge disappeared.

Faraday set a copper disc to rotate between the two poles of a horseshoe ferro-magnet, with one sliding contact in touch with the edge of the disc and another fixed to the pivot on which the disc rotated. As the disc rotated a measurable current of electricity flowed from the sliding contact to the fixed one. When the disc stopped rotating the electricity disappeared.

This was the first crude generator of electricity which over the subsequent 100 years was developed to what are now large complex generators that still depend on the principle of rotating a conductor through the field of force around a magnet, to generate electricity as an alternating current as the conductor rotates through the north and south polarity of the magnet.

ELECTRICAL SUPPLY

Supply generators

In England and Wales the bulk generation of electricity is carried out by PowerGen, National Power and Nuclear Electric who distribute power through the national grid to twelve regional electrical companies (RECs) and also directly to large consumers of electricity. In Scotland Scottish Hydro-Electric and Scottish Power generate and distribute electricity.

National grid

The national grid distributes electricity at 132 000 volts (132 kV), mainly by overhead conductors. A high voltage is used to minimise transmission losses in the great length of cable by the use of an economic section of copper conductor.

Supply from the national grid is converted by regional companies to lower voltages of 11 000 and 6600 volts, to supply districts within their areas, and this is further reduced to the standard 415 volts for local supply. The electricity supply to the majority of users is the standard 415 volt, three phase 50 Hz frequency alternating current, supplied through a cable consisting of three phase wires and a neutral conductor.

Over a period of some 100 years of development, a supply of electricity, alternating at a frequency of 50 Hz, has been accepted as the most efficient, economic and convenient. The windings of the generator are arranged so that the supply of alternating currents is phased, with three pulses of energy each revolution of the generator – the supply best suited to most electrically driven motors.

Low voltage supply

The regional electrical companies make a connection to the majority of buildings through a low voltage, 415 volt, three phase supply cable comprising three separate phase wires and one neutral conductor. The neutral conductor serves as a return path to complete the circuit for current flow back to the generator.

Where the anticipated load is low or moderate and where rotary equipment is used, connection to the consumer is made to any two of the three phase wires and the neutral to provide a 415 volt, three phase alternating supply. Where the anticipated load is low, as for a house, connection for the consumer is made to any one of the three phase wires and the neutral conductor to provide a 240 volt single phase supply.

High voltage supply

Low voltage is defined as exceeding 50 volts and not exceeding 1000 volts a.c., and high voltage as exceeding 1000 volts a.c.

Where the anticipated load on the supply is high, as for example to heavy industry or a common supply is made to a large building, it is necessary for the regional electricity company to provide a high voltage supply. This need is usually determined by the particular regional electricity company's policy on loading. The general division between low and high voltage intakes lies between 250 to 500 kVA.

The high voltage supply is connected to the REC's switch gear, from which a connection to the consumer's switch gear is made to the consumer's chosen supply system of one or more distribution centres.

Tariffs and metering

The majority of consumers are supplied by the RECs, who offer the option of three tariffs for the payment for supply:

(1) General purpose tariff
(2) Low voltage maximum demand tariff
(3) High voltage maximum demand tariff

General purpose tariff

This tariff is for relatively small consumers such as domestic premises. The tariff consists of a quarterly standing charge plus a charge per unit consumed, applicable to lighting, heating and power. In some regions a reduced charge for off-peak consumption, during night-time or weekends, may be available, which requires separate meters and circuits.

Low voltage maximum demand tariff

The purpose of maximum demand tariffs is to penalise consumers who exceed a maximum demand, agreed between the supplier and the consumer, in order to provide a reasonably uniform level of demand on the supply available and so avoid unpredictable demand.

Where the consumer exceeds the agreed maximum demand for any one month, they are required to pay a higher charge, based on consumption for that month, for the following 12 months. Were they to reduce consumption in any month after that year, they would not revert to the former lower payment until twelve months later. There is thus a two year period of higher payment for exceeding demand for one month.

Because it is very costly to shut down and start up generators to meet surges in demand, and it is impractical to reduce output, tariffs are designed to ensure steady demand as far as possible. To this end monthly tariffs generally vary monthly, being lowest from March to the end of October and higher from November to the end of February – the months when demand is highest. The charge consists of a monthly unit based on the actual kWh metered, with units metered during night-time being considerably lower than daytime, giving a reduction of up to 50%.

High voltage maximum demand tariff

This tariff is similar to the low voltage tariff in that it penalises consumers who exceed the agreed maximum demand and rewards those who take supplies in the summer months and during the night and at weekends.

Cost control

The penalties for exceeding agreed maximum demand of electrical supply may be considerable. It is a matter of economic good sense, therefore, for business to reduce peak demand where possible. Where electrically powered rotary machinery is used, there is a peak demand first thing in the morning when machinery is started up simultaneously. A phased starting, either organised for manual operation or by automatic means, can produce considerable financial reward with very little interruption to production.

Metering

The electricity supply authority has an obligation to provide a meter to record basic data, on which a tariff charge is based. The majority of small meters in use are of the induction type which record units consumed, to which a tariff is applied to calculate the charge to the consumer. The induction meter, which may be liable to electrical and mechanical failure, should be serviced and must be withdrawn at intervals for testing to comply with legal requirements.

Electronic meters are available and are being installed in some areas. It is likely that electronic meters will gradually replace induction meters. The advantage of the electronic meter is that remote reading of the meter is possible for computer recording at central points.

Electrical circuits

To effect the transfer of energy from a source of potential electrical power, such as a generator or transformer, a complete circuit of some conductive material is necessary to provide a path of low resistance back to the source, so that the maximum energy is available around the circuit for conversion to heat for lighting and heating and for motive power to rotary equipment. The material most used as a conductor is metal in the form of a copper or aluminium wire.

Phase conductors

With alternating current (a.c.) three phase supplies – the usual electrical source – there are three separate conductors, one to each of the three separate phase windings of the generator or voltage reduction transformer, and a fourth conductor serving as neutral back to the source to complete the circuit.

The three separate wires, which are termed phase or phase wires, are sometimes called live or live wires.

Neutral

The conductor which serves to complete the circuit is termed neutral.

Earth

In addition to the phase and neutral conductors, there is another separate conductor termed earth which may be a separate conductor in some cables or for economy may be combined in other cables with the neutral conductor. Fig. 178 shows the terms used.

The earth conductor serves as a protective and safety device by acting as part of a conductive circuit with earth, which is said to be neutral electrically. The earth conductor serves as a line of least resistance to the discharge of current in excess of that allowed for in the design of an electrical installation, which might otherwise damage equipment, cause a fire or endanger life.

The earth conductor to supply cables, which is for the benefit of the suppliers' equipment, may not be an effective earthing conductor for a consumer's installation. As the supplier is under no obligation to provide satisfactory earthing for the consumer, it is up to consumers to provide their own earthing provisions as necessary.

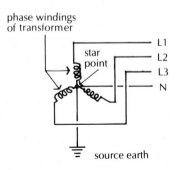

Fig. 178 Phase conductors (L1, L2, L3) and neutral for TN-C-S supply.

Where the three phase supply is used as a low voltage, low demand supply to, for example, a single house, a single phase connection is made to one of the three phases wires and the neutral of the supply cable for a 240 volt supply. If the incoming supply cable uses a combined neutral and earth conductor, as is usual, the consumer's single phase supply will need a separate earth conductor.

Main distribution

The electrical supply necessary for large buildings, groups of buildings and manufacturing industry is run to the site by the REC, through a transformer substation as necessary, to their main switch, circuit breaker equipment and meter. From the meter it is the responsibility of consumers to install their own main distribution system of cables to supply transformers and switch gear at load points on extensive sites, or some form of busbar distribution system for large buildings such as multi-storey buildings and factories.

A busbar is a round or rectangular copper or aluminium bar conductor, made in a range of standard lengths. The busbars are either of bare copper, which is supported at intervals by insulated carriers, or the copper bars can be totally insulated. The advantage of the bare copper busbar is that sub or final circuits can readily be connected through tee-offs, by clamping to rectangular bars at any point. The round section bar requires shaped connections.

Busbars are run inside solid enclosures or galvanised steel trunking in which there are insulating supports. Tap-off boxes, complete with miniature circuit breakers, can be fed to the trunking as necessary for sub or final circuits. For long runs, for example to supply fixed position lighting, the busbar will generally be PVC insulated single-core cable supported inside metal trunking fixed to the underside of a floor or roof, with socket outlets for individual lights. For heavier loads on horizontal runs, the insulated busbar will connect to terminals with plug-in units, complete with circuit breakers, to connecting machine cable feeds.

Vertical risers of busbar distribution systems are usually run as bare rods supported by insulated carriers inside metal trunking from which fused tap-off points are connected to each floor for connection to final circuits. Vertical risers that are not inside a fire compartment must be provided with insulated barriers at each floor as fire stops.

Fig. 179 shows busbar distribution to a block of flats, feeding distribution boards, circuit breakers and final circuits to floors.

Main distribution for high voltage supplies is run as either a radial or ring main system. Radial circuits run from main switchboard to an outlet and back to the main switchboard, with a circuit breaker for each radial feeder. This is the simplest and cheapest form of circuit for

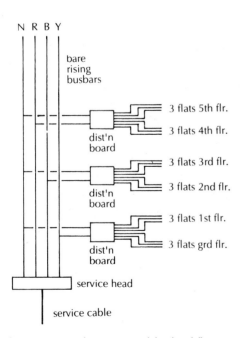

Fig. 179 Distribution to a block of flats.

high voltage main distribution, which is used for one or a few supply outlets. Ring main circuits run to a number of tap-off points and back to the main switchboard, with circuit breakers to each end of the ring system and switchgear for tapping-off to radial feeders or transformers.

The advantage of a ring circuit is that it supplies a number of outlets, additional outlets can be tapped-off without shutdowns, and maintenance on the feeder circuits is facilitated.

Final circuit distribution

Final circuits are those that feed directly to lighting or power fittings from the mains. In small buildings, such as houses and shops, the final circuits run from one mains distribution board fixed close to the entry point of the mains supply. In larger buildings the final circuits run from single tap-off points on radial or ring main circuits or from distribution boards connected to radial, ring or busbar mains distribution. The distinction between main and final circuits is in the anticipated loads, which are heavier on the main than final circuits and affect the size of the necessary cable, switchgear and circuit breakers for the circuits.

For the sake of economy in the size of cable, switchgear and circuit breakers, final circuits are divided into various circuits depending on the anticipated loads, and in larger installations are divided into nonessential and essential services.

Final circuits for houses, shops and small offices are run separately for:

(1) Lighting from fixed ceiling and wall fittings
(2) Portable fittings such as air heaters, kitchen equipment and portable lamps
(3) Fixed equipment such as cookers and water heaters.

For larger buildings, final circuits are run separately for nonessential services such as:

(1) Lighting, fixed light fittings
(2) Small power fittings such as heaters and other portable equipment
(3) Lifts
(4) Air conditioning
(5) General plant and fixed equipment

and for essential services such as:

(1) Critical processes such as computers
(2) Security systems
(3) Emergency lighting

(4)   Fire alarms
(5)   Communications

Low voltage supply

Low voltage three phase supplies by the RECs to consumers' premises are usually provided through cables that comprise three separate phase conductors. Each is insulated with a cross-linked polyethylene (XLPE) insulating covering around which a combined neutral and earth conductor, in the form of a metal sheath, is embedded in unvulcanised rubber surrounding the insulated phase conductors, and the whole is encased in a protective PVC sheath.

Earthing systems

TN–C–System

The earthing arrangement which is combined in a common neutral and earth conductor is referred to as the TN–C–S system. The letter T denotes the connection of the star point of the three phase supply. N denotes the connection of exposed conductive parts, such as metal pipes, with the source earth that is earthed neutral. C–S denotes the protective conductor combined in the supply cable and separate in the consumer installation. This system is also known as the pme system (protective multiple earthing).

TT system

Another earthing system used is the TT system where the second T denotes that the exposed conductive parts are connected by independent installation earth electrodes, by protective conductor to source earth. This system was used for overhead power supplies.

Another system, the TN–S, employs separate neutral and earthed protective conductors. This older system is less used.

Supply connections

The work of connecting the incoming 415 volt, three phase supply cable to the consumer's electrical installation is carried out by the REC's engineer after the electrical installation is completed and tested. For a 240 volt supply to small premises a connection is made to one of the three phase wires through the REC's main cut-out fuse housed in a sealed box. The combined neutral and earth metal sheath conductor of the supply cable is connected to an earth connection block provided by the consumer. From the REC's main cut-out fuse and earthing block, short lengths of single core insulated cable are connected to the REC's meter.

Meter tails

The consumer's electrical engineer will have installed a consumer unit in some position close to the anticipated entry point of the supply cable. Connected to the consumer's unit will be two short lengths of single wire – insulated cable ready for the REC's engineer to connect to their meter. These cables are commonly termed 'meter tails', and should be as short as practical – no more than 3 m long. Because the overcurrent protection of these tails is provided by the REC's main cut-out fuse, they require a minimum cross sectional area of the wire to these cables to be rated to the capacity of their fuse. For single phase supplies a minimum cross sectional area of conductor to each tail is specified – 10, 16 and 25 mm^2 for main cut-out fuses of 60, 80 and 100 amp rating respectively. For security it is good practice to use the largest cross section area of cable to allow for fuse replacement errors.

Fig. 180 shows connections from a supply cable to a single phase consumer unit.

It is common today for the supplier's meter and cut-out fuse to be installed inside a metal meter case fixed externally to facilitate meter reading. The meter and fuse are housed in a meter box that should preferably be set in a recess in an external wall to provide some protection from rain. Meter tails are run from the consumer's unit, fixed internally, to the meter.

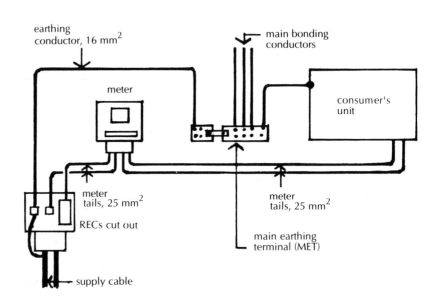

Fig. 180 Supply connections for a TN–C–S system.

ELECTRICAL DISTRIBUTION

Consumer's installation

A consumer's electrical installation for small premises, such as a house, begins at the connection of the meter tails to the REC's meter. From the meter a low voltage single phase, 240 volt supply is run to the consumer's installation, which includes a consumer's unit and the necessary separate circuits for lighting, heating and power.

A consumer's unit combines in one factory-made unit the necessary functions of a switch for isolation, circuit breaking devices (fuses) and distribution of supply to the various circuits. An earthing connection block is fixed adjacent to the consumer's unit. Fig. 181 shows the components of a consumer's unit.

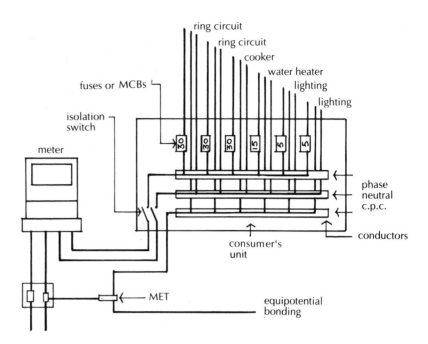

Fig. 181 Connections to consumer's unit and circuit distribution.

Isolation main switch

Isolation is the word used to describe the function of a device that effectively cuts off all voltages to the whole of an electrical installation or to a complete circuit, for the purpose of work to the installation, in case of fire and when premises are left unoccupied. For a single phase supply a double pole switch is included in the consumer's unit for safety, to make a break in both the phase (live) and neutral conductors by the operation of a manual switch.

Distribution

The three distribution conductors housed in the consumer's unit are connected separately to the phase and neutral of the supply and to the earth conductor made through the earth connection block. From the three distribution conductors the supply to each separate circuit is run from the phase conductor through an overcurrent protective device (fuse) and directly from the neutral and earth conductor to each circuit.

Main earthing terminal (MET)

To provide a means of connecting the earthing conductor of the TN–C–S supply cable to the means of earthing for the consumer's installation and to the main equipotential bonding, a main earthing terminal (MET) is included as part of the consumer's installation. The MET is made as two blocks of metal with a disconnectable link, as illustrated in Fig. 180. The disconnectable link is provided as a means of testing.

A conductor is run from the supplier's common earth and neutral conductor to the smaller MET block, and a conductor is run from the main MET block to the earth distribution in the consumer's unit. This provides the necessary earth connection to the various circuit protective conductors (cpcs), commonly called earth, to protect the installation against damage and danger of fire or shock to persons. Where there is a failure of insulation to a live conductor, and contact between the conductor and a conductive part of the installation, the circuit protective conductor provides an alternative path of least resistance for unpredicted currents to flow to earth.

One or more earth conductors are run from the main MET block to the extraneous conductors, such as metal water, gas, oil, and heating pipes, as main equipotential bonding. This bonding is shown as separate conductors in Fig. 186.

Overcurrent protective devices

In the phase (live) conductor to each circuit cable run from the distribution conductor, is a fuse or circuit breaker as protection against current greater than that which the circuit can tolerate. The purpose of these devices is to cause a break in a circuit as protection against damage to conductors and insulation by overheating caused by excessive currents.

Fuses

The fuse is the original form of overcurrent protective device which operates through a thin wire that is designed to overheat and rupture at a pre-determined maximum current, and so break the circuit. The three types of fuse in use are:

(1) Semi-enclosed rewirable fuse
(2) Cartridge fuse
(3) High breaking capacity (hbc) cartridge fuse

Semi-enclosed rewirable fuse

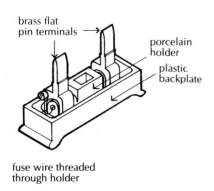

brass flat
pin terminals

porcelain
holder

plastic
backplate

fuse wire threaded
through holder

Fig. 182 Rewirable fuse.

The semi-enclosed, rewirable fuse (Fig. 182) consists of a porcelain fuse holder through which the fuse wire is threaded and connected to the two brass terminals. The fuse is pushed into position in the fuse block to make electrical contact. When the fuse wire has ruptured or blown, the holder is pulled out, a new fuse wire fitted and the fuse pushed back into place. This, the cheapest fuse available, has lost favour to the cartridge fuse and the circuit breaker which are easier to replace or reset.

The disadvantages of the semi-enclosed fuse is that the wrong type of wire may be fitted, and in time the wire may oxidise and not function as intended. Fuse wire is usually rated at 5, 10, 15 and 30 amp.

Cartridge fuse

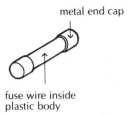

metal end cap

fuse wire inside
plastic body

Fig. 183 Cartridge fuse.

A cartridge fuse consists of a fuse wire in a tube with metal end caps to which the wire is connected. The fuse wire is surrounded by closely-packed granular filler. When a circuit overcurrent occurs the wire heats, ruptures and breaks the circuit, the energy released being absorbed by the granular filler without damage to the fuse carrier. Fig. 183 shows a cartridge fuse.

These fuses are cheap, easy to replace by pressing into place between terminals and do not deteriorate over time. They are extensively used as circuit breakers in plug tops to provide protection to the flexible cable and moveable equipment connected to socket outlets. Cartridge fuses are made with ratings from 2 to 15 amps.

High breaking capacity fuse

The hbc (high breaking capacity) cartridge fuse consists of a ceramic tube with brass end caps and copper connecting tags to which the silver elements inside the tube are connected. The elements are surrounded by granulated silica filling to absorb the heat generated when the elements overheat, rupture and break circuit. These more sophisticated and expensive fuses are used for the more heavily loaded installations.

Circuit breakers

A circuit breaker is a thermal-magnetic, magnetic-hydraulic or assisted bi-metal tripping mechanism designed to operate on overload to break the connected circuit.

Miniature circuit breakers (MCBs)

Miniature circuit breakers (MCBs) are extensively used as protection against damage or danger resulting from current overload and short circuit in final circuits in buildings.

The simplest form of MCB consists of a sealed tube filled with silicon fluid in which is a closely fitting iron slug. When overload occurs the magnetic pull of the charged coil surrounding the tube causes the iron slug to move through the tube and trip the circuit breaker switch, which closes. To test or make the circuit it is only necessary to open the switch, which will remain open if the cause of the overload has been removed or will close if it has not.

The advantage of these circuit breakers is that there is no wire or cartridge to replace and the operation of a switch is all that is needed.

More sophisticated MCBs depend on thermo-magnetic trip operation. MCBs are factory moulded, sealed units available with terminals for plugging in or bolting to metal conductors in consumers' units. A range of ratings is provided to suit the necessary overload current ratings for particular final circuits.

The Wiring Regulations recommend disconnection times for overcurrent protective devices of 4 sec for connections to fixed equipment and 5 sec for portable and hand-held equipment.

Final circuits

Final circuits are the circuits of a consumer's installation that complete a circuit for the flow of current back to the supply neutral. For the low voltage, single phase alternating current usual for small premises, a small compact consumer's unit will provide distribution terminals for the number of separate final circuits used. Typical circuits for a house could be one or more ring main circuits and one cooker circuit, each with a 30 amp breaking capacity fuse or MCB and two or more lighting circuits each with a 5 amp fuse or MCB.

The purpose of segregating the various circuits is to afford economy in the cost of cables, the cross section of those with 5 amp fuses being smaller than those with a 30 amp rating. Fig. 181 shows a typical consumer's unit layout.

The length of each circuit is limited to provide an economical section of cable and to minimise the electrical resistance of the cable and the number of connections to each circuit, to limit overload current.

The two types of circuit that are used are the ring and the radial.

Ring circuit

Ring circuits are commonly used for socket outlets that provide electrical supply to portable equipment such as vacuum cleaners, electric fires, portable lamps and kitchen equipment, through a socket outlet and a plug top. The ring circuit (Fig. 184) makes a big loop or ring from one outlet to the next, round all the outlets and back to the consumer's unit, as the most economic and convenient cable layout.

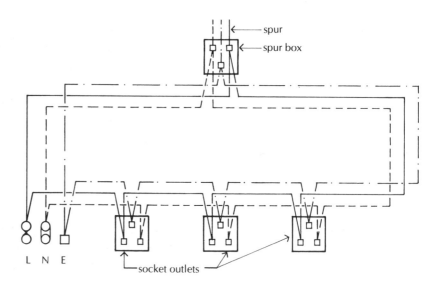

Fig. 184 Ring circuit.

The recommended maximum length of cable run depends on the cross section area of conductor chosen and the type of protective device used. While there is no recommended limit to the number of outlets served, it is recommended that a maximum of $100\,\text{m}^2$ of floor area be served by a ring circuit, protected by a 30 amp fuse or MCB in domestic premises.

The cable generally used for ring circuits is $2.5\,\text{m}^2$ conductor, twin and earth, PVC insulated and PVC sheathed cable.

Spur outlets

Where a socket outlet cannot be conveniently fed by a ring circuit a spur outlet may be run from the ring circuit. A spur outlet is connected through a joint box or spur box, as shown in Fig. 184. In effect the spur outlet is a radial circuit run off a ring circuit. There is no limit to the number of spurs which can be connected to a ring circuit other than a limit that there should be no more spur outlets than there are outlets fed directly off the ring. The cable to the spurs should be the same as that for the ring circuit.

Radial circuits

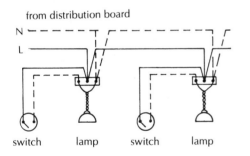

from distribution board

switch   lamp   switch   lamp

Fig. 185 Radial lighting circuit.

The circuit arrangement for lighting is generally in the form of radial circuits, each of which runs from the consumer's unit to its light fittings and back to the consumer's unit, as if radiating out. In each radial circuit the phase conductor runs in the form of a loop from the circuit through a switch back to the light fitting or fittings that the switch controls (Fig. 185). (For the sake of clarity the earth conductor is not shown in Fig 185.) By this arrangement the switch controls the fittings allocated to it without controlling the rest of the light fittings connected to the circuit.

The radial circuit is adopted for individually switch controlled light fittings as the most convenient and economical means of running cables. Where one switch is used to control several light fittings the circuit may be run as a ring circuit off a radial circuit, or as a ring circuit by itself, whichever is the most convenient.

A 5 amp fuse or MCB is connected as overcurrent protective device to the phase wire run out from the consumer's unit. The cables are run in $1.5\,\text{mm}^2$ conductor, twin and earth, PVC insulated and PVC sheathed cable.

To limit current flow to suit the fuse or MCB and cable size, there are usually two or more separate lighting circuits to most small installations such as those for a house.

Safety requirements

For safety the Wiring Regulations (BS 7671) recommend good workmanship in the running of cables to avoid damage to insulation and conductors, and the making of connections and sound judgement

in the selection of materials to prevent danger from shock to persons and the possibility of fires from overheating of conductors due to current overload by short circuit.

Short circuit

A short circuit is, as the words imply, a fault in a circuit where a live conductor comes into contact with another, by breakdown of insulation to provide a path of least resistance to flow with consequent greater current than allowed for in the circuit design. A short circuit may cause overheating of conductors and breakdown of insulation and so damage the installation. It is to limit such occurrences that overcurrent protective devices, such as fuses and MCBs, are fitted.

Earthing

The Wiring Regulations (BS 7671) include recommendations for protection against the dangers of electric shock to persons or livestock, damage to installations and the danger of fire from overcurrents and earth faults. Earth faults occur when a live conductor makes contact with a conductive part of the installation, or extraneous conductive parts such as metal service pipes, and current flows to the neutral mass of earth. The earth conductor in cables and the earthing connections to conductive metals provide an alternative path for unplanned flows of electrical energy.

Electric shock

Electric shock to persons is differentiated as direct contact and indirect contact.

Direct contact

Shock from direct contact is caused when a person comes into contact with a live electrically charged part which causes a current to flow through them to earth. This is likely to cause injury which may be fatal.

It is generally accepted that a voltage over 50 may be fatal. Such heavy current flows, which are in effect short circuits, will, within 4 or 5 secs, cause protective devices to break circuit. Direct contact with a properly designed installation is unlikely.

Indirect contact

Indirect contact occurs when contact is made with an exposed conductive part of an electrical installation, such as the metal casing to an electric fire, which is not normally live but may have become so under earth fault conditions caused by contact of a live conductor with a metal casing due to insulation failure. The third conductor, the circuit protective conductor (cpc), is included in circuits and connected to metal casings to electric fires to serve as a conductor to earth for such unpredicted currents.

With the generally used TN–C–S system of supply cables, the earth (cpc) conductor of an installation is connected to the combined neutral earth of the supply cable, which may not provide a wholly satisfactory path to earth. It is generally necessary, therefore, to provide other earth paths through earth electrodes connected to the installation's earthing block.

It was practice for many years to make a connection to earth through a metal service pipe entering the building from underground. With the now common use of plastic pipework for service pipes, such as gas and water, this is no longer accepted as a satisfactory earthing arrangement.

Main equipotential bonding

The Wiring Regulations (BS 7671) require that all extraneous conductive metal parts, which are not part of an electrical installation, be connected to the main earthing terminal (MET). This earth bonding is provided as protection against the possibility of conductive metal outside (extraneous) the electrical installation becoming live due, for example, to failure of insulation of a cable run close to a heating pipe, making a live connection to the pipe. The conductive metals included are water, gas, oil, heating and hot water pipes, radiators, air conditioning and ventilation ducts, which should be earth bonded as illustrated in Fig. 186.

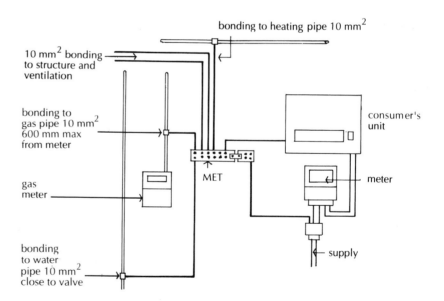

Fig. 186 Equipotential bonding.

Cross bonding

The purpose of equipotential bonding is to co-ordinate the characteristics of protective devices with earthing and the impedance (resistance) of the circuit to limit touch voltages until the circuit protective devices cause disconnection.

Because of the introduction of plastic pipes, connections and plastic coatings to metal, it may be necessary to provide earth bonding connections across plastic used as part of a whole system, where good sense indicates the possibility of indirect contact being made to metal that might become live.

Supplementary equipotential bonding

It is a requirement of the Wiring Regulations (BS 7671) that supplementary equipotential bonding be provided to all simultaneously accessible, exposed conductive parts in special conditions of wet activities and high humidity, such as bathrooms and swimming pools, where the moisture in such conditions provides additional risks due to the conductivity of water.

The term 'simultaneously accessible' in relation to exposed conductive parts means the possibility of someone, for example in a bath of water, reaching out to touch a heated, metal towel rail and so providing a conductive path from the metal bath to the towel rail, through their body. If the towel rail and its pipe connections had become electrically charged, due to a breakdown of insulation to a cable in contact with the pipes, there would be a possibility of shock to the individual. The purpose of supplementary bondings is to spread the voltage potential across and between other adjacent bonded metal parts to equalise and so limit the voltage potential to that least likely to cause injury or death in the few seconds before circuit breakers come into operation.

A bath, in which someone is immersed in water, is considered the appliance most requiring supplementary bonding between metal pipes and a metal waste pipe, as shown in Fig. 187. Bonding to a basin is required between metal hot and cold water pipes and a metal waste pipe. Similarly a metal supply pipe to the cistern of a WC should be bonded to a metal waste pipe.

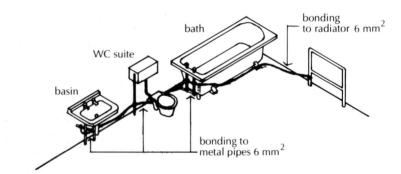

Fig. 187 Supplementary equipotential bonding.

To complete the supplementary bonding between all simultaneously accessible exposed conductive parts, there should be bonding between the cross bonding to a bath, basin, WC and a radiator. This connecting bonding may be effected by a conductive pipe that provides a link to sanitary fittings. Where there is no satisfactory conductive link by pipework, conductive bonding should be connected to all bonded fittings and connected to the radiator as a terminal.

To effect supplementary bonding a conductor, preferably of copper, should be connected between simultaneously accessible, exposed conductive parts by means of pipe clamps, through which the conductor should run continuously. The bonding conductor may be of bare copper or insulated copper, the lowest cross sectional area of which should be $4\,mm^2$.

Because it is a requirement that supplementary bonds be accessible for inspection and testing, there may well be some untidy exposed conductors and pipe clamps which cannot be behind removable bath panels or basin pedestals.

Supplementary bonding is only necessary where there are exposed and simultaneously accessible metal parts. It would be futile to bond to plastic pipework and covered metal pipework.

CABLES AND CONDUITS

Cable for final circuits

VIR cable

Vulcanised india rubber (VIR) cable was extensively used for final circuits before PVC cable was first produced. The VIR cable consists of an inner layer of rubber around tinned, copper wires with an outer coating of vulcanised rubber. The rubber outer coating of this cable becomes brittle with age and fails, requiring replacement after some 20 years. It was often run inside metal conduit from which old cable could be withdrawn and new cable pulled through. This type of cable has largely been replaced by PVC cable.

PVC cable

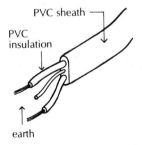

Fig. 188 PVC insulated, PVC sheathed cable.

The cable most used today for wiring to final circuits is PVC cable, either as PVC insulated, PVC sheathed cable or PVC insulated cable. In the former the phase and neutral copper wires are each separately covered with PVC as insulation and then all are surrounded with a PVC protective sheath in which the earth (cpc) copper conductor is enclosed, as shown in Fig. 188. The outer PVC sheath is for protection against damage during installation. PVC insulated cable comprises a copper conductor wire insulated with PVC. This is for use when the cable is protected by the conduit in which it is run.

PVC (polyvinyl chloride) is a tough, incombustible, chemically inert plastic which is an effective insulator that does not deteriorate during the useful life of most installations.

PVC twin and earth

The cable most used for final circuits to 240 volt, single phase supplies is PVC insulated, PVC sheathed cable commonly described as PVC twin and earth, describing the phase and neutral insulated conductors and the earth in the sheath. The size of the cable is defined by the cross sectional area of the copper wires – 1.5 mm² or 2.5 mm² for lighting and socket outlet circuits respectively.

For 415 volt, three phase supplies a three core PVC insulated and sheathed cable is used.

Protection

The Wiring Regulations (BS 7671) provide extensive recommendations on measures to be taken to minimise damage to cable during installation and in use. These may be broadly grouped under the headings mechanical damage, temperature, water and materials in contact including corrosive materials.

Mechanical damage includes precautions to avoid damage during installation, such as drilling holes in the centre of the depth of floor joist for cable runs rather than using notches in the top or bottom of joists where subsequent nailing might damage cable.

At comparatively high and low temperatures plastic may soften or become brittle respectively, and so weaken. At temperatures above 70°C plastic will appreciably soften, and below freezing will become noticeably brittle and crack.

Water, both as liquid and in vapour form, in contact particularly with terminals, may act as conductor between live and live and live and earth conductors.

To protect cables from damage by corrosive materials such as cement, cables run in plaster, concrete and floor screeds should be protected by channels or conduit.

Mineral insulated metal sheathed cable

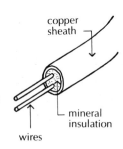

copper
sheath

mineral
insulation

wires

Fig. 189 Mineral insulated copper sheathed cable.

This type of cable consists of single-stranded wires tightly compressed in magnesium oxide granules, enclosed in a seamless metal sheath of copper or aluminium, as shown in Fig. 189. The combination of the excellent insulation of the magnesium oxide and the metal sheath gives this cable an indefinite life and reasonable resistance to mechanical damage. For added protection the metal sheath may be protected with a PVC sheath.

Because of the high initial cost of the cable and the various fittings and seals necessary at bends, junctions and terminations, the use of this cable is confined to commercial and industrial installations where the cost may be justified. The metal sheath may serve as the earth (cpc) conductor.

Channels

Channels of plastic or galvanised steel, generally in the form of a hat section, are used to provide PVC cable, which is to be buried in plaster, with protection and to secure the cable in position while the plaster is being spread. The brim of the hat section is tacked to the wall surface around the cable. These sections which are used solely as protection do not provide a ready means of pulling cable through when replacing cable.

Conduit

Round or oval section plastic or galvanised steel sections are used as protection where PVC cable is to be run in plaster, solid floors, walls and roofs. Conduit is used to provide all round protection, particularly where the cheaper PVC insulated cable is used, and to provide a means of renewing and replacing cable by drawing cable through the conduit. Because of the bulk of the conduit it is generally necessary to form or cut chases (grooves) in wall surfaces and solid floors and roofs where screeds are insufficiently thick to accommodate the conduits.

Plastic conduit

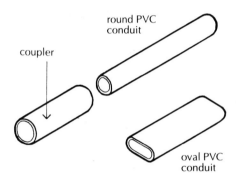

Fig. 190 PVC conduit.

Plastic conduit, which is considerably cheaper than metal conduit, is made in round or oval sections (Fig. 190). Round section conduit which can be buried in thick screeds has to be fixed in a chase or groove in walls. Oval section conduit can be buried in thick plaster finishes and thin screeds.

Lengths of conduit are joined by couplers, such as the one shown in Fig. 190, into which conduit ends fit. A limited range of elbows and junctions is made for solvent welding to conduit. The conduit is secured with clips that are tacked in position. There is some facility for pulling through replacement cable, but not as much as with metal conduit.

Because it is made from plastic, the conduit will not serve as earth conductor, as does metal conduit.

Metal conduit

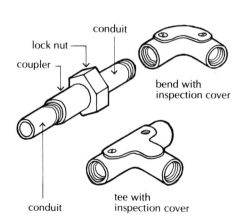

Fig. 191 Steel conduit.

Metal conduit is manufactured as steel tubes, couplings and bends which are either coated with black enamel or galvanised. The cheaper black enamel conduit is used for cables run in hollow floors and roofs and other dry situations. The more expensive galvanised conduit is used where it is buried in concrete floors, roofs and screeds and for chases in walls below plaster finishes where wet finishes might cause rusting if black enamel conduit were used.

Fig. 191 shows a steel conduit. The nominal bore of the conduit is its outside diameter, a range of sizes being produced.

Light and heavy gauge steel conduit is made, light gauge with push-fit connections and heavy gauge with screwed connections. The cheaper light gauge conduit does not provide as positive conductive connections as the heavy gauge, as earth conductor. Heavy gauge conduit is for use in concrete floors, walls, and roofs, where it provides protection against damage during the process of placing and compacting wet concrete.

Metal conduit is produced with a variety of fittings designed to facilitate pulling out old cable and replacing it with new. The access or inspection covers to the bend and tee, shown in Fig. 191, are for pulling through.

Because the conduit is of conductive metal it may serve as the earth (cpc) conductor, and single PVC insulated cable may be used as the conduit provides protection against damage.

Trunking

Where there are extensive electrical installations and several comparatively heavy cables follow similar routes, with standard conduit too small to provide protection, it is practice to run cables inside metal or plastic trunking to provide both protection and support for the cables.

Trunking is fixed and supported horizontally or vertically to wall, floor or ceiling structures, or above false suspended ceilings, with purpose-made straps secured to the structure. Trunking is run inside ducts for vertical runs and above false ceilings where appearance is a consideration or exposed in commercial premises. Trunking may be hollow, square, rectangular, circular or oval in section. Square or rectangular sections are usually preferred for convenience in fixing to walls or ceilings.

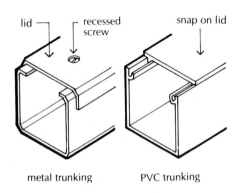

lid — recessed screw   snap on lid

metal trunking   PVC trunking

Fig. 192 Metal trunking.

Fig. 192 is an illustration of square trunking with access for inspection and any necessary renewal or alterations. A range of square and rectangular trunking sizes is made, together with elbows, tees and gusset fittings. Large trunking systems are usually purpose-made from galvanised steel sheet to suit particular needs.

OUTLETS

Socket outlets

Socket outlets are so called because they comprise sockets in a wall-mounted front plate into which the terminals of a loose plug top fit to supply electricity to a wide range of moveable, portable and hand held electrical equipment.

Socket outlets are usually connected to a ring final circuit, there being no limit to the number of sockets connected to each ring circuit, which should not serve an area greater than $100 \, \text{m}^2$ for domestic installations. A ring circuit is protected by a 30 amp fuse or MCB at the consumer unit.

A socket outlet consists of a galvanised steel box to which a front plate is screwed after the terminals of the front plate have been connected to the electric supply cable. The galvanised, pressed steel boxes are made with circular knockouts, which can be removed for cable entry, and lugs for screws to secure the plastic front plate. Steel boxes are usually made for recessing into a wall and plaster and others for surface fixings.

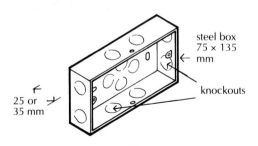

steel box 75 × 135 mm

knockouts

25 or 35 mm

Fig. 193 2 gang steel box.

Fig. 193 shows a recessed steel box for a two gang, two plug top outlet. Steel boxes for surface fixing have a smooth faced finish with the knockouts for cable entry and holes for fixing screws in the back of the box. Fig. 194 shows a steel box for surface mounting.

Front plates to socket outlets are moulded from plastic to suit single, double or multi-gang plug top outlets, complete with sockets for square three pin plug tops. Brass terminals for the phase, neutral and earth conductor cables are fixed to the back of the front plate, and it is holed for two screws for fixing to the steel box. Socket outlets may be supplied with a switch and a pilot light to indicate the 'on' position. Fig. 195 shows a single outlet front plate, with optional switch and pilot light and a plug top.

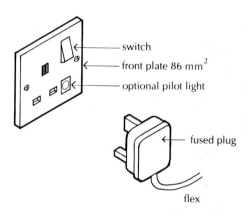

Fig. 195 13 amp outlet and plug.

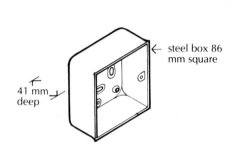

Fig. 194 Steel box for surface mounting.

Many socket outlets are connected to a ring circuit which is protected with a 30 amp fuse or MCB at the consumer unit. It is necessary to fit a cartridge fuse to the plug top to provide current overload protection to the flexible cord and appliance connected to the outlet. Fuses to plug tops to outlets vary from 2 or 4 amps for lighting to 13 amps for electric fires.

Where there is a spur branch to a ring circuit a spur box is fitted to provide a connection for the spur cables and to protect and control the spur outlet. A spur box consists of a steel box and plastic front plate in which is a fuse and a switch (Fig. 196). The terminals at the back of the front plate serve to make connection of the spur cables.

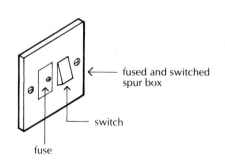

Fig. 196 Spur box.

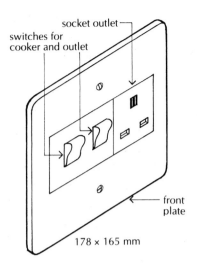

Fig. 197 Cooker control unit.

To suit the wide range of electrical equipment in use today, multi-gang socket outlets are common, particularly in kitchens and where electronic equipment is located. Multi-point socket outlet units with up to six outlets are available for fixing to kitchen worktops, test benches, laboratories and computer desks. These units, which are connected to a ring circuit outlet, can be fixed in position. They can be moved to another location if need be.

The electrical supply to a cooker with hobs, grill and oven requires a high fuse rating and cables. A separate radial circuit is run from the consumer's unit to the electric cooker, and protected by a 30 amp fuse or MCB in the consumer's unit. A wall mounted cooker control unit is fixed close to the cooker. This control unit consists of a recessed steel box and plastic front plate in which a single switch or two switches and a socket outlet are provided, as shown in Fig. 197. The socket outlet in the cooker control unit is for such portable equipment as an electric kettle. The two switches control the cooker and the socket outlet.

Lighting outlets

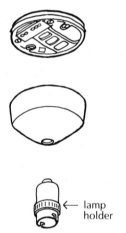

Fig. 198 Ceiling rose and lampholder.

Lighting outlets may be connected to a radial circuit for ceiling and wall lights, where one or a few lighting fittings, recently termed luminaires, are controlled by a wall switch, or they may take the form of socket outlets for portable table lamps, connected to a ring circuit or one or more radial circuits depending on the number of outlets and convenience in wiring.

The traditional ceiling light, which was common for high ceilings, consisted of a ceiling rose screwed to the ceiling, from which a pendant cable dropped supporting a lamp holder fixed at some convenient height. The ceiling rose was connected to a radial circuit protected by a 5 amp fuse in the consumer unit, with a loop down to a wall switch which controlled the ceiling light. Fig. 198 shows traditional ceiling light with ceiling rose and lampholder.

Because a pendant lamp is unsuited to the lower ceiling height of most modern domestic buildings, a batten lampholder may be used for single ceiling lamps or one of the many lighting fittings designed to fit closely to ceilings.

The batten lampholder shown in Fig. 199 is assembled as a single unit with the lampholder close to the ceiling. The three cables, phase, neutral and earth, are connected to the terminals in the base, which is screwed to the ceiling. The lampholder is screwed to the base and the lampholder cover screwed around the lampholder.

Lighting outlets may be connected to a radial circuit for ceiling and wall lights where one or a few fittings are controlled by a wall switch, or a ring circuit where several portable lights, such as table lamps, are plugged into socket outlets and controlled by a common switch.

Fig. 199 Batten lampholder.

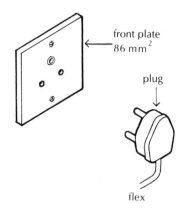

Fig. 200 2 amp outlet and plug.

Fig. 200 shows a front plate and plug top for a lighting outlet for portable lamps. The face plate is screwed to a steel box set in plaster finish just above floor level. The three pin plug top is connected to the flexible cable to a light fitting.

The conventional ceiling and table lamp illumination, by traditional incandescent filament lamps, is still much favoured as domestic lighting for the so called 'soft effect' of such shaded light, which does not show sharp contrast of light and shade.

Of recent years the more intense lighting systems used for display purposes in shops and showrooms have been used in the home for kitchen and general lighting. Systems of 'downlighting' have been used, consisting of a number of lights recessed into ceilings and systems of 'spotlighting', using lamps fixed to tracks in the ceiling, which may be adjusted to light a particular area.

Wall switch

A wall switch is generally recessed in a wall, partition or plaster depth for appearance sake. A galvanised, pressed steel box is set into a recess in the wall after knockouts have been removed for the cable entry. The three conductors, phase, neutral and earth, are connected to the terminals set into the back of the plastic front plate, which is then screwed to lugs in the steel box.

Fig. 201 shows a single gang switch, with steel box and plastic front plate.

For small rooms a single switch inside the access door or opening is generally sufficient for ceiling and wall lighting. For large area rooms it is often convenient and economical to use two or more circuits, each with its own wall switch, so that a part of, or the whole of, the room may be illuminated.

For staircases a system of two-way or three-way switching to a circuit is used to control the lighting on upper and lower floors as necessary. The system of two way or three way switching involves

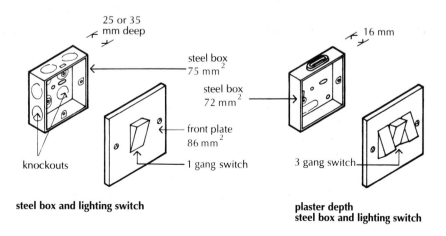

Fig. 201 Wall switches.

additional wiring to provide the means of switching from two places. Fig. 201 is an illustration of a steel box and plastic front plate for a three gang switch.

Clock connector box

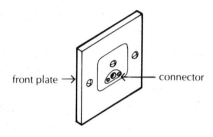

Fig. 202 Fused clock connector box.

Wall mounted electrically operated clocks which consume little electricity are usually connected to a radial lighting circuit which is protected by a 5 amp fuse or MCB at the consumer unit. The clock is connected through a fused connector box which comprises a steel box and plastic front plate. The connector and fuse are housed in a separate plate screwed to the front plate (Fig. 202).

7: Gas Supply

GAS AS A FUEL

Town gas

Town gas, first produced in 1812, is the combination of combustible gases from the carbonisation (heating) of coal. Each town had its own gasworks and gasholders supplying gas for lighting, cooking and heating – hence the name town gas. A byproduct of the heating or carbonisation of coal to produce town gas was coke, which was extensively used as a cheap fuel for heating.

With the introduction of electricity the demand for gas declined. During the first half of the twentieth century there was keen competition between the suppliers of gas and electricity to the advantage of the consumer. The advantage of a lighting and heating source at a competitive rate at the touch of a switch led to the changeover from gas to electricity, first for lighting and later for domestic heating. The manufacture of town gas required a large site, a ready source of high quality coal and a considerable labour force, all of which were becoming increasingly scarce and expensive.

Natural gas

Natural gas, first imported from North Africa in 1964 and supplied to consumers has since 1968 come from the North Sea fields which, it is estimated, have known reserves sufficient for current and future consumption well into the twenty-first century.

Natural gas, which is mainly methane, has twice the calorific value of town gas and is a high grade controllable fuel eminently suited to both domestic and commercial uses for heating. Since the introduction of natural gas, its use by industry and commerce has increased fourfold.

Unlike town gas, natural gas is nontoxic and as it is odourless, additives give warning of leaks by their distinctive smell. Natural gas is delivered to the consumer at pressures two to three times greater than town gas and in consequence pipes about half the bore of those needed for town gas can be used.

Since 1968 most town gas burning appliances have been converted to use natural gas and all new appliances are designed to burn natural gas. The conversion of appliances was effectively completed in 1976.

Combustion of gas

For the ignition and subsequent combustion of natural gas (methane), a supply of oxygen and a gas temperature of about 700°C is necessary.

In gas-fired cookers, fires and boilers the necessary oxygen is taken from an intake of air which is drawn into the combustion chamber or zone, mixes with gas, is ignited and burns to complete combustion.

By-products of combustion

The products of the complete combustion of gas are carbon dioxide and water vapour which are expelled, due to the pressure of combustion and the intake of air, to outside air by flues or by convection.

Lethal carbon monoxide

Where there is an insufficient intake of air to provide oxygen for the complete combustion of gas, the by-product will contain carbon monoxide, a gas that in very small quantities is lethal and can cause death in a few minutes.

Conditions for optimum combustion

The design of gas burning appliances is concerned primarily, therefore, to provide an optimum mix of gas (methane) and air for complete combustion for both efficiency in the use of fuel, safety, adequate convection currents for the intake of air and the evacuation of the by-products of combustion.

GAS SUPPLY

Service pipe

Gas is supplied under pressure through the gas main from which a branch service pipe is run underground to buildings. The service pipe is laid to fall towards the main so that condensate runs back to the main, where it is collected in buckets.

Where the consumption of gas is high, as in commercial and other large buildings, a valve is fitted to the service pipe just inside the boundary of the site to give the supplier control of the supply, for example in the case of fire. Domestic service pipes are run directly into the building without a valve at the boundary, and the meter valve or cock controls the supply.

The gas service pipe must not enter a building under the foundation of a wall or loadbearing partition, to avoid the possibility of damage to the pipe by settlement of the foundations. Gas service pipes running through walls and solid floors must pass through a sleeve so that settlement or movement does not damage the pipe. The sleeve is usually cut from a length of steel or plastic pipe larger than the service pipe which is bedded in mastic to make a watertight joint.

Gas meter installation

The service pipe connects to the supply pipe through a gas meter installation which comprises a cock or valve governor, filter and a meter, for domestic premises, with the addition of a thermal cut-out and non-return valve for larger installations.

Where possible, domestic meter installations are outside the premises in a position affording shelter, such as in a basement area under steps, or in a box or housing giving shelter. The advantage of fitting the meter outside is that it is naturally ventilated and in a position where the meter reader can gain access when the occupier is out.

Meter installations to large premises are often in a purpose-built meter house.

Gas cock (valve)

A gas cock to control the supply from the service pipe to the governor and meter consists of a solid plug that in the shut position fills the bore of the cock, and in the open position only partly obstructs the flow of gas. The gas cock is operated by a hand lever, as illustrated in Fig. 203. When the lever is in line with the service pipe the cock is open, and when it is at right angles to the pipe it is closed.

The connection of the gas cock to the pressure governor is made with a short length of semi-rigid stainless steel tube which can accommodate any movement between the service pipe and meter which might otherwise damage the meter. The semi-rigid tube is illustrated in Fig. 203.

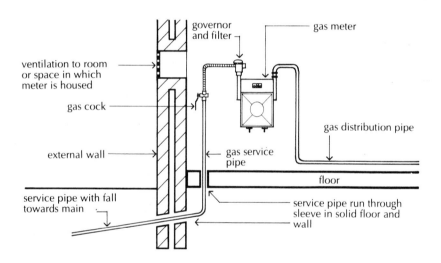

Fig. 203 Gas supply.

Pressure governor and filter

For domestic installations a combined pressure governor and filter is fitted at the connection of the service pipe to the meter, as illustrated in Fig. 203, and a separate governor and filter are used for larger installations. The fine mesh filter is fitted to collect pulverised particles of rust and metal which are carried along the main by gas. But for the filter these fine particles might clog gas jets to gas burning appliances.

The governor is a spring-loaded diaphragm valve, the function of which is to reduce the pressure of gas in the main to a pressure suited to gas burning appliances. The governor reduces mains gas pressure to 20 to 25 mbar standing pressure.

Gas meter

The meter illustrated in Fig. 204 is the traditional tin case gas meter. The light gauge tin case contains bellows that fill with gas through a valve and then discharge gas to the distribution pipe so that the movement of filling and emptying operates the meter that records the volume of gas supplied. As the meter records the volume of gas supplied it should not be near to a heat source otherwise the consumer will be paying for the heated and therefore greater volume of gas. Because of the flimsy construction of the tin case meter, it is practice to make connections of service and distribution pipes with either a semi-rigid stainless steel tube or a short length of lead pipe which can take up any movement and so prevent damage to the meter. Semi-rigid and lead pipe connections are illustrated in Fig. 204.

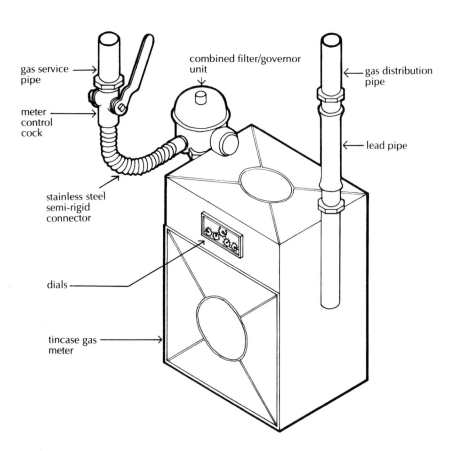

Fig. 204 Domestic gas meter installation.

Because both the pressure at which it is delivered and the calorific value of natural gas are higher than that of town gas, an appreciably smaller meter is used. These rigid steel case meters which contain compact bellows may be up to half the size of the old tin case meter illustrated in Fig. 204. Because of the greater rigidity of the steel case it is practice to make the service pipe connection with a semi-rigid tube and the distribution pipe with a rigid connection.

Where gas consumption is large a rotary displacement meter is used.

Meter box

It is practice today to fit gas meters in some position on an outside wall where there is access for meter reading. A metal meter box is fixed either on an external wall face or fitted to a recess in a wall to give some protection from weather. In this position the meter control cock or valve is more available for access in emergencies such as fire, than it is indoors.

Where interruption of the gas supply for maintenance, repair or replacement of the meter or governor is unacceptable, as in hospitals and some industries, it is usual to install a meter bypass. A bypass is a length of pipe connected directly between the service pipe and the distribution pipe to bypass the meter. There should be a separate gas cock, governor and filter in the bypass so that the meter cock may be closed for maintenance or repair work and the bypass cock opened meantime to continue the supply.

The room or space in which a gas meter is installed should be permanently ventilated to the open air against escape of gas and build up of heat. A small air brick or vent suffices for most domestic installations and ventilation equivalent to 4% of the floor area for larger installations.

PIPEWORK

The gas pipes which are run from the meter to supply the various appliances may be described as supply pipes. Commonly the system of pipes is described as a gas carcass and the work of running the pipes as gas carcassing.

The traditional pipework for gas carcassing for town gas was mild steel tubulars with a natural steel finish, as galvanised coatings are adversely affected by gas. Because of the lower calorific value of town gas and the larger volume of gas required, large bore pipes were used.

Today, because of the smaller bore of pipes required and labour saving convenience, copper tubulars are used for most small gas installations. Capillary or compression fittings at connections are used in the same way that water pipework is run.

Pipe sizes

For safety the size or bore of pipe used for gas carcassing should be adequate to deliver the necessary pressure of gas to each appliance. If the pipe is undersize, too low a pressure of gas at appliances may result in incomplete combustion and development of carbon monoxide, or there may be a flashback into pipework and an explosion may occur.

The required size of pipe from the meter and the branches to each fitting, such as cooker, boiler and gas fire, depends on the volume and pressure necessary and the frictional resistance of the pipes and fittings to flow.

The necessary sizing of pipework in a particular gas carcassing installation can be assisted by a simple line isometric diagram similar to that described for water pipe calculations, where dimensions of pipe lengths are shown and each identified by a letter. From this the actual length and additional length allowance for fittings is determined, to give an effective length and resistance to flow. A possible pipe size is assumed and checked against pressure loss, to provide adequate supply to appliances. The procedure for gas pipe sizing is similar to that for sizing water pipes.

Testing pipework for leaks

Newly installed gas carcassing may be tested for leaks before the pipework is connected to appliances. This involves capping off open ends of pipe. More usually pipework is tested after the gas burning appliances are connected and their supply shut off by closing gas cocks on each appliance. The pipework may be tested either before or after the meter has been installed.

The procedure for testing after the meter has been installed is to turn off the main gas cock and all gas appliances and pilot lights. The screw of the test nipple, on the outlet of the meter, is removed and a pressure gauge or manometer is connected to the nipple. The main gas cock is slowly opened to let gas into the pipework until a standing pressure of 20 to 25 mbar is indicated on the manometer. The gas is turned off at the main cock, and after a wait of one minute, over the next two minutes the pressure indicated on the manometer should show no drop in pressure greater than 4 mbar, for domestic installations.

If the pressure drops below the 4 mbar limit there is a leak, which can be indicated by leak detection liquid which, when sprayed around a leak, will bubble to indicate the position of the leak.

Purging

Purging is the operation of evacuating all air which has entered when the pipelines are first installed and when the gas is turned off for maintenance and repairs.

The purpose of purging gas pipelines of air is to avoid the possibility of gas mixing with air in the pipelines. Were this to occur and the jets to appliances ignited, the mixture of air and gas in the pipeline might also ignite and cause an explosion.

Air is purged by opening the main control cock and the cock to an appliance to cause the gas entering under pressure to force out all air. The gas is allowed to flow until there is a distinct smell of gas when the appliance or burner cock is closed. During this operation windows are opened and naked flames and electric sparks are avoided until the escaped gas is dispersed.

FLUES

Flues to gas burning appliances

The by-product of the combustion (burning) of gas and air is heated gases that will rise by natural convection. The purpose of a flue or ventilation system is to encourage the heated gases to rise, by convection or by force of a fan to outside air. Where the heated gases do not have a ready, adequate escape route to outside, they will mix with inside air.

On mixing with inside air the water vapour in the gases will cool to form condensation on cold surfaces, and the interruption to steady convection currents may cause incomplete combustion of gas and the mixing of small amounts of poisonous carbon monoxide gas with inside air.

Open flue appliances

Open flue appliances depend for their operation on an intake of air directly from the room or enclosure in which they are fixed. To this end there must be a permanent opening to outside air through which adequate air may be drawn to replace that drawn in by the appliance.

Ventilation

Where the permanent ventilation does not allow sufficient replacement air to enter the room, inefficient combustion may occur due to reduction of air pressure in the room and that may cause discomfort to the occupants and possible entry of poisonous carbon monoxide gas.

The inclusion of effective draught seals to windows and doors in modern buildings severely restricts the entry of adequate outside air to replace that required for combustion where there is insufficient positive permanent ventilation.

Natural convection flues

Natural convection flues depend for their operation on the natural draught or pull of heated gases rising from heating appliances. Heated gas expands and rises naturally to replace cooler more dense air. It is the function of flues to encourage heated gases to rise vertically from heating appliances to outside air. In general the higher a flue rises the greater the draught or pull of hot gases will be. Plainly a flue should rise directly with as little change of direction from the vertical as is practical.

Size of flue

There is an optimum size of flue to provide the best draught. The best cross sectional dimensions of a flue depend on the output from a heating appliance and the volume of air required for combustion of a particular appliance.

Draught diverter

A vigorous draught of gases up a flue is encouraged by draught plates and hoods over open flued boilers, gas fires and open fires. These draught diverting plates and hoods are designed to direct the heated combustion gases up the flue and to divert unwanted draughts of air away from the combustion chamber.

Existing brick or block built flues

Where gas burning appliances are fitted in existing buildings it is common to use an existing open fire brick flue as the flue to the new appliance. The flue is first swept to clear loose, friable material such as soot and broken brick or mortar, and the draught up the flue is tested with a burning oily rag or a smoke producing device to test the draught caused by the intake of naturally rising air. This will test for both the draught up the chimney flue and for an adequate intake of air by natural ventilation of air into the room.

Where flues are built into an external brick or block wall it may be necessary to line the flue to minimise condensation of flue gases inside the flue, caused by too rapid a cooling of the rising gases.

The condensation of moisture vapour from flue gases inside brick or block flues may cause unacceptable staining of wall surfaces and damage to chimneys, where the expansion of water soluble crystals in brick faces exposed to cold winds may adversely affect the structure.

Flue liners

To provide the best section of unobstructed flue inside an existing brick or block chimney it may be necessary to line the flue with a flexible stainless steel liner which is pulled up the flue and sealed to a flue terminal and the gas appliance at the base.

Flues built into new brick or block buildings are lined with clay pipes or purpose made flue blocks.

Flue pipes

Instead of flueways, flue pipes may be used. The flue pipe to a heating appliance may be independent of, or secured to, a wall. Stainless steel, enamelled steel or aluminium are used for both single walled and double walled pipes.

Double walled flue pipes consist of two concentric pipes separated by an insulating material. This insulating material maintains the heat and convection draught of flue gases and prevents the pipe becoming too hot, where it is exposed.

Flue terminal

Where a flue rises to open air it is finished with some form of terminal formed to deflect gusts of outside air that might otherwise blow down flues and temporarily block escape of flue gases.

If a flue is to be effective in discharging flue gases to open air it should rise above adjacent roofs and structures so that air currents blowing towards the flue terminal do not cause down draughts or suction currents as the blown wind gusts are deflected by surrounding buildings.

Room sealed appliances

Room sealed gas burning appliances are those that draw their combustion air intake directly from outside air instead of from inside air, as is the case with open flue appliances.

The majority of small capacity boilers used for space and water heating for domestic premises such as flats and houses are specifically designed to be small and compact to fit into kitchen or bathroom. These so called 'space-saving' boilers operate through a terminal fixed to the external face of a wall, into which air is drawn and from which combustion gases escape either by natural convection or fan assisted operation. With natural convection they rely on natural air movements and use a balanced flue to control intake and expulsion of gases. The other type use an electrically operated fan to assist extraction of combustion gases.

All these appliances are sealed so that no part of the intake air or exhaust gases enter the room in which the appliance is fixed. In operation both the balanced flue terminal and the smaller fan draught terminal will adequately disperse flue gases with extension to flues or flue pipes.

8: Refuse Storage

REFUSE COLLECTION

Refuse volume

The volume of loosely packed refuse from an average three person household is $0.09\,m^2$ per week, slightly less than the capacity of a standard $0.092\,m^2$ dustbin. The larger part of the volume of domestic refuse today is bulky lightweight paper and plastic wrapping and container material which, in our 'throw away' style of living, is increasing to the extent that it is estimated that the volume of refuse from an average three person household will increase to $0.12\,m^2$ before long.

This bulky refuse encourages the householder to compress as much as possible into his refuse bin, which makes it difficult to discharge the contents into the refuse collection vehicle. The bin is damaged by banging it against the vehicle to empty it and in a short time it too becomes refuse and the cycle of waste accelerates. The smaller part of domestic refuse is ash from solid fuel appliances, tins, bottles and kitchen waste. The latter, if not wrapped, may adhere to the side of the bin, putrefy and be the source of disagreeable odours and a breeding ground for flies.

The usual sequence in the storage of domestic refuse is the filling of a small bin or other small receptacle inside the dwelling, which is emptied into a refuse bin, a larger refuse storage container or into a refuse chute discharging to a refuse container.

Collection is usually once or twice a week. Refuse is collected from the premises or from the kerbside, depending on access and local arrangements, and larger containers are usually collected from the premises by vehicles designed for the purpose.

The required capacity of refuse containers depends on an assumption of about $0.3\,m^2$ refuse per person and the frequency of collection. It is sensible to provide a larger capacity than this average to cover interruption of collection during holiday periods and festivals when the volume of bulky, lightweight refuse increases considerably; otherwise refuse bins will be packed and difficult to empty or will have inadequate capacity with resulting spillage of refuse. An additional capacity of 10–25% is not unusual.

DUSTBINS (REFUSE BINS)

Refuse bins are still generally described as dustbins from the days when the solid fuel fire was the principal source of heat and the resultant volume of dust and ash that was discharged to bins gave them their name. At that time a more frugal style of living did not produce the volume of refuse common today.

Galvanised mild steel dustbin

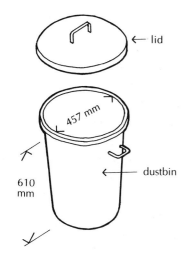

Fig. 205 Standard galvanised steel dustbin.

This is the traditional dust or refuse bin that has only recently been superseded by the plastic bin and the paper sack. A soundly made galvanised steel dustbin is robust and will give useful service for many years providing the zinc coating (galvanising) is not damaged by mishandling. Once the zinc coating wears, the mild steel rapidly rusts and the bin disintegrates.

The standard mild steel dustbin illustrated in Fig. 205 is round in section and tapers from top to bottom to facilitate emptying and also stacking. It has a reinforcing turnover rim, slightly dished bottom, lifting handles and a loose lid. The capacity of standard bins is $0.028\,\text{m}^2$ (1 ft), $0.056\,\text{m}^2$ (2 ft), $0.071\,\text{m}^2$ ($2\frac{1}{2}$ ft) and $0.092\,\text{m}^2$ ($3\frac{1}{4}$ ft). The $0.071\,\text{m}^2$ and $0.092\,\text{m}^2$ bins are those most used. A standard steel bin is heavy, about 12 kg, and when full it is near the limit in weight that can be lifted and emptied by an average person without strain.

Various non-standard light section galvanised steel bins are manufactured, mostly from corrugated or fluted sheet to reinforce the flimsy material. Because of the light section material from which they are made they have an appreciably shorter life than the heavier standard bin, and refuse that collects in the troughs of the corrugated or fluted sides is difficult to clean. These bins, though somewhat cheaper, are a false economy.

Rubber lids and rubber bases to steel bins are available to reduce the noise of handling the bins.

A specialised-steel bin, the dustless loading bin, is manufactured for the storage and emptying of ash and other dusty refuse. The lid is hinged to the bin and so designed that it does not open until the bin has been lifted by the special collection vehicle and sealed against a shutter for dustless emptying. This type of bin is too heavy for manhandling and has to be wheeled on a trolley to the collection vehicle.

Plastic refuse bins

Plastic refuse bins are about half the weight of a standard steel bin of the same capacity and if made of high density polyethylene or polypropylene are rigid, durable and may have a useful life of several years if reasonably handled. They do not deteriorate by oxidisation as do steel bins. There is no great difference in cost between the standard steel and the plastic bin. A good quality plastic bin, such as that illustrated in Fig. 206, has taper sides without flutes or corrugations, a reinforcing rim, lifting handles, and a loose lid. Usual capacities are $0.071\,\text{m}^2$ and $0.092\,\text{m}^2$.

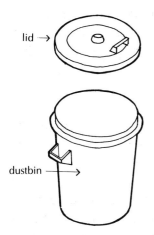

Fig. 206 Plastic dustbin.

These bins do tend to deteriorate fairly rapidly when manhandled in being emptied into collection vehicles. Being lightweight they are liable, if free standing, to be blown about in high winds when empty.

Lightweight, low density polythene plastic bins are manufactured with corrugated sides for reinforced. These flimsy bins are brittle and easily fractured, particularly at low temperatures, and do not have a reasonable useful life.

Square section plastic bins on wheels are supplied by some local authorities to reduce the labour of collection. These comparatively small, so-called 'wheelybins' are often overfilled and difficult to empty and so defeat the object of their design, which is to reduce labour.

Paper and plastic refuse sacks

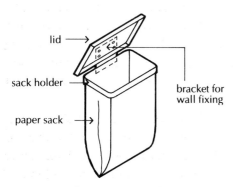

Fig. 207 Wall mounted holder and refuse paper sack.

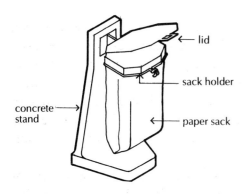

Fig. 208 Free standing holder and refuse paper sack.

First used in Scandinavia, the disposable paper refuse sack is an alternative to the steel or plastic bin for householders. The obvious advantage is that both the soiled container and its contents are collected and disposed of in one journey from and to the refuse collection vehicle. But the natural resource, wood, from which sacks are made is becoming increasingly scarce and expensive in our throwaway society and it seems unlikely, despite its advantages, that the paper sack will replace the steel and plastic bin.

Paper sacks for refuse are made from stout two ply wet strength paper or single ply waterproof kraft paper. These sacks are sufficiently robust to stand outside between normal collections and to store all but the most jagged of items of refuse without damage. Bags of capacity of $0.071\,\mathrm{m^2}$ and $0.092\,\mathrm{m^2}$ are generally used. The refuse sacks are supported by wall mounted or free standing holders, as shown in Figs. 207 and 208. Wall mounted holders are fixed to a wall with a back plate which supports the sack holder and its lid. Alternatively, free standing concrete or metal stands support sack holder and lid. For collection the full sack is unclipped and replaced with a fresh one. Free standing holders are heavy enough to stand in high wind and remain upright against knocks. Both wall mounted and free standing sack holders may be fitted with wire guards to protect against damage.

Plastic sacks have been used instead of paper sacks; they are cheaper than paper, require less space for storage of sacks and do not deteriorate in damp conditions. Plastic sacks are fixed to wall mounted or free standing holders similar to those for paper sacks.

To date the majority of paper and plastic refuse sacks have been supplied by local refuse collection authorities as a part of their refuse collection service, as a manpower saving device where the bulk of the refuse is collected from individual households.

This rational system of refuse storage depends on a degree of careful and sensible use by the householder. The all too frequent abuse of this system by overfilling and forcing in angular objects very soon reduces the efficiency of the system by causing needless spillage of refuse and consequent extra labour on the part of the collectors or the householder, or both.

REFUSE CHUTE

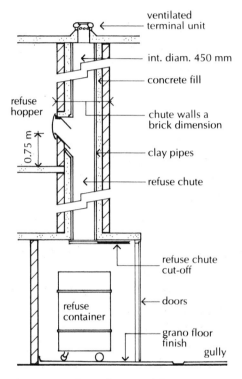

ventilated terminal unit

int. diam. 450 mm

concrete fill

refuse hopper

chute walls a brick dimension

clay pipes

0.75 m

refuse chute

refuse chute cut-off

doors

refuse container

grano floor finish

gully

Fig. 209 Refuse chute and container chamber.

For domestic buildings of more than four floors, a system of refuse chutes is a sensible means of disposal and storage. A refuse chute is a vertical shaft into which refuse is tipped through hoppers, the refuse being collected and stored in a container at the foot of the chute. Refuse chutes are lined with cylindrical pipes of clay, concrete or fibre cement, internal diameter not less than 450 mm, the smooth impervious surface of the pipes providing the least impediment to the movement of the refuse down the chute and facilitating cleansing by periodic hosing down. The lining pipes are enclosed in a brick or concrete shaft for their support and as a protection against spread of fire.

Metal hoppers at each floor level provide entry points to the chute. The opening to these hoppers should not exceed 350 mm in width and 250 in depth. Hoppers to chutes should be located on open communal access balconies or well ventilated lobbies away from habitable rooms or in separate well ventilated lobbies of fire resisting construction, off main circulation lobbies.

At the foot of each chute there should be a container chamber as illustrated in Fig. 209. When the refuse container is full, the chute is closed by the steel shutter, the full container replaced with an empty one, and the shutter opened. Depending on the anticipated volume of refuse a variety of arrangements for containers is available, such as the single 0.95 m^2 wheeled container illustrated in Fig. 210, a range of single containers on a turntable, a range of refuse sacks on a turntable or a single large container.

Refuse chutes should be carried up to or above roof level with a ventilating terminal of the same diameter as the chute, or where this is not possible with a reduced ventilating pipe and terminal. Ventilation of the chute and lobbies in which the hoppers are located is essential.

With sensible use, reasonable periodic changes of container to avoid spillage, and cleansing of the chute and container chamber, the refuse chute is a satisfactory system for storage of refuse in multi-storey buildings.

The disadvantage of these chutes is that they can be somewhat noisy in use when heavy objects fall from higher hopper entries. Due to thoughtless use they become blocked with large cardboard boxes and such things as umbrellas which are difficult to clear.

WASTE DISPOSAL SYSTEMS

Refuse containers

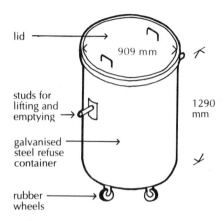

lid

909 mm

1290 mm

studs for lifting and emptying →

galvanised steel refuse container →

rubber wheels →

Fig. 2.10 Galvanised mild steel refuse storage container.

The Garchey system

Refuse containers are large metal containers in which refuse, both domestic and trade, is stored. The limit to the size of these containers is the capacity of a collection vehicle to lift and carry or tow away the container. The galvanised steel container shown in Fig. 210 is often described as a Paladin container.

These containers are wheeled for manhandling to the collection vehicle which is designed to lift, upturn and empty the contents into the rear of the collection vehicle. To this end there are various lifting attachments to the container such as the studs shown in Fig. 210, or angle iron rims to suit the various makes of collection vehicle. These standard $0.95 \, m^2$ refuse containers are extensively used at the foot of chutes for communal and trade refuse. A wide variety of large, purpose-constructed, galvanised steel containers are available, principally for storage at the foot of chutes and for trade waste.

These heavy containers are somewhat difficult to manhandle towards the collection vehicle and are noisy in the operation of mechanically lifting them into the vehicle for emptying. They are rarely cleaned due to the difficulty of access to the inside of the container and are liable to become smelly in warm weather.

The Garchey system is a method of waste disposal in which refuse is fed through an enlarged waste outlet in the sink into a waste tube housed inside a refuse receiver, fitted below the sink. Waste water from the sink runs into and fills the waste receiver, as illustrated in Fig. 211. When the waste tube is filled it is raised by the householder and its contents are washed down the waste to the 150 mm waste stack to the collection chamber. All waste water appliances are connected to the waste stack so that their discharges assist in washing down the refuse. Soil appliances are drained to a separate stack.

The Garchey refuse collection chamber is emptied once a week to a tanker which removes the waste from the refuse, and carts it away. Surplus water is drained to the sewer. Large material such as papers and contains has to be broken down before being fed into the system.

This refuse disposal system has not been extensively used because of its high initial cost and high maintenance due to careless usage. It was introduced some years ago for use in multi-storey blocks of flats, and is suited to the disposal of wet and damp waste such as that from kitchens. It is not, however, suited to the disposal of the increasing bulk of plastic containers, plastic film and paper wrapping that is common today.

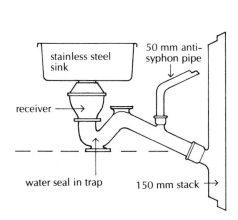

Fig. 211 Garchey refuse disposal unit.

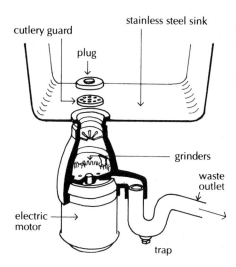

Fig. 212 Waste disposal unit.

Sink waste disposal units

Kitchen waste is fed through the sink waste to a disposal unit in which a grinder, powered by a small electric motor, reduces the refuse to small particles that are washed down with the waste water from the sink. These units are designed to dispose of such kitchen refuse as food remains which rot and cause disagreeable odours in bins. They are not suited to the disposal of larger bulky lightweight refuse. Fig. 212 is an illustration of one of these units.

In common with other seemingly sensible innovations in refuse disposal, this system has lost favour largely because of the impatience of users and the need for fairly frequent maintenance.

Bin liners

Of recent years plastic bin liners have become one of the most used ways of containing refuse for disposal. These cheap liners are convenient for both impatient, careless householders and for refuse collection operatives, but with the disadvantage of unsightly black bin liners littering the streets, and inevitable spillage.

Index